TRIUMPH OR TRIAGE? THE WORLD FOOD PROBLEM IN GEOGRAPHICAL PERSPECTIVE

C. Gregory Knight
R. Paul Wilcox
The Pennsylvania State University

RESOURCE PAPER NO. 75-3

1710 Sixteenth Street, N.W.
Washington, D.C. 20009

Library of Congress Card Number 76-029265
ISBN 0-89291-115-8
Supported by the AAG and by a grant from
the National Science Foundation

FOREWORD

In 1968, the Commission on College Geography of the Association of American Geographers published its first Resource Paper, *Theories of Urban Location*, by Brian J. L. Berry. In 1974, coinciding with the termination of NSF funding for the Commission, Resource Paper number 28 appeared, *The Underdevelopment and Modernization of the Third World*, by Anthony R. deSouza and Philip W. Porter. Of the many CCG activities, the Resource Papers Series became an effective means for permitting both teachers and students to keep abreast of developments in the field.

Because of the popularity and usefulness of the Resource Papers, the AAG applied for and received a modest grant from NSF to continue to produce Resource Papers and to put the series on a self-supporting basis. The present Resource Papers Panel subscribes to the original purposes of the Series, which are quoted below:

> The Resource Papers have been developed as expository documents for the use of both the student and the instructor. They are experimental in that they are designed to supplement existing texts and to fill a gap between significant research in American geography and readily accessible materials. The papers are concerned with important concepts or topics in modern geography and focus on one of three general themes: geographic theory; policy implications; or contemporary social relevance. They are designed to implement a variety of undergraduate college geography courses at the introductory and advanced level.

In an effort to increase the utility of these papers, the Panel has attempted to be particularly sensitive to the currency of materials for undergraduate geography courses and to the writing style of these papers.

The Resource Papers are developed, printed, and distributed under the auspices of the Association of American Geographers, with partial funding from a National Science Foundation grant. The ideas presented in these papers do not imply endorsement by the AAG.

Many individuals have assisted in producing these Resource Papers, and we wish to acknowledge those who assisted the Panel in reviewing the authors' prospectuses, in reading and commenting on the various drafts, and in making helpful suggestions. The Panel also acknowledges the perceptive suggestions and editorial assistance of Jane F. Castner of the AAG Central Office.

Salvatore J. Natoli
Educational Affairs Director
Association of American Geographers
Project Director and Editor, Resource Papers Series

Resource Papers Panel:

John F. Lounsbury, Arizona State University
Mark S. Monmonier, Syracuse University
Harold A. Winters, Michigan State University

PREFACE

In writing *Triumph or Triage* we have come to conclusions that have left us personally uncomfortable, conclusions that we only vaguely suspected when we began the project. The reader may share our discomfort if our conclusions are seen to be valid. We doubt that this paper can be read without reaction, however strongly one may differ from our perspective. The world food situation provides an excellent opportunity to examine not only the problem itself but also the role of science and scientists in society. At the very minimum, we are all participants in the world food scene.

We will be using this resource paper principally in our courses on human use of the environment. Here, the paper will focus discussion of the nature of food supply systems toward a problem-solving perspective. For lower division students, the paper offers considerable data and concepts for consideration and elaboration. For advanced students, it is a point of departure toward specific issues and methodological ideas. For all students, food and famine pose persistent questions with a vast and accumulating literature awaiting scrutiny from a geographer's viewpoint.

One fundamental problem in addressing such a diffuse topic is finding specific foci for further study and reflection. The paper sacrifices depth for breadth in order to provide a multiplicity of starting points. In preparing the paper we found that our preconceived perspectives (such as adaptive systems and the concept of vulnerability) were insufficient to allow us to address the range of questions implied in the world food problem; we hope the student can benefit from the considerable investment we made before other foci began to emerge.

An additional problem is the identification of data sources. We have tapped only the surface of a vast amount of data and literature, and our citations will suggest sources from which one can develop more detailed analyses. There are innumerable data sets on food production and consumption for manipulation, mapping, and analysis. The apparently prosaic nature of mapping, tabulation, or statistical analysis is deceiving. In working with data, one invariably discovers questions that might otherwise have been missed and is forced to recognize both the virtues and the limitations of data being used.

The reader will quickly discern major topics relevant to the world food problem that we have oversimplified or completely ignored. One of these is the disparate nature of food consumption at a subnational level as differentiated by age, sex, income, ethnic identity, region, or other characteristics. Another is the food production and provision system of centrally planned or socialist economies. In addition, we have bypassed the historical processes by which industrial agriculture evolved, as well as the enticing and important living historical farms such as Old Sturbridge Village in Massachusetts and Living History Farms in Iowa. There is no end to meaningful directions for individual and classroom exploration.

We trust that this paper will prove useful in other than specialized courses. Virtually any dimension of basic human geography can be illustrated by questions of food and famine, and it is in introductory courses that this paper may have its greatest impact. A seminar on the world food problem, though, sounds intriguing . . .

C. Gregory Knight
R. Paul Wilcox
The Pennsylvania State University

CONTENTS

LIST OF FIGURES

LIST OF TABLES

The Blind Men and the Elephant

It was six men of Indostan
To learning much inclined,
Who went to see the Elephant
(Though all of them were blind).
That each by observation
Might satisfy his mind.

The First approached the Elephant,
And happening to fall
Against his broad and sturdy side,
At once began to bawl:
"God bless me! but the Elephant
Is very like a wall!"

The Second, feeling of the tusk,
Cried, "Ho! what have we here
So very round and smooth and sharp?
To me 'tis mighty clear
This wonder of an Elephant
Is very like a spear!"

The Third approached the animal,
And happening to take
The squirming trunk within his hands,
Thus boldly up and spake:
"I see," quoth he, "the Elephant
Is very like a snake!"

The Fourth reached out his eager hand,
And felt about the knee.
"What most this wondrous beast is like
Is mighty plain," quoth he;
" 'Tis clear enough the Elephant
Is very like a tree!"

The Fifth who chanced to touch the ear,
Said: "E'en the blindest man
Can tell what this resembles most;
Deny the fact who can,
This marvel of an Elephant
Is very like a fan!"

The Sixth no sooner had begun
About the beast to grope,
Than, seizing on the swinging tail
That fell within his scope,
"I see," quoth he, "the Elephant
Is very like a rope!"

And so these men of Indostan
Disputed loud and long,
Each in his own opinion
Exceeding stiff and strong.
Though each was partly in the right
And all were in the wrong!

John Godfrey Saxe (1882)

INTRODUCTION

Problems of food supply and famine are among the most bewildering, diffuse, and frustrating of mankind's contemporary dilemmas. Within the lifetime of each of us, official views of the world food situation have ranged from dire predictions of starving hordes to expectations of a nirvana of plentiful food supply, only to return again to impending doom. Even while one view was in vogue, there remained proponents of the opposite opinion. One expert states that famine is imminent; another that our ability to feed the world's people adequately is finally within reach. As residents of an industrialized nation, we are told that advanced agricultural technology not only provides us with food for a lower proportion of income than elsewhere, but is also the hope for successful agricultural development in the poorer nations. On the other hand, we are accused of profligate consumption of resources and wasteful adherence to diets that are morally unjustifiable in a hungry world. The fundamentals of our political and economic systems, along with our technology, are questioned in the face of persistent hunger within the industrialized nations. Even among those who agree that there is a world food problem, there is widespread disagreement on its causes and potential consequences, let alone its solutions. Commodity food aid, the diffusion of technology, and policies of triage are among the responses that have been proposed.[1] It is understandable that we may see ourselves as one of Saxe's blind men, when even experts cannot agree among themselves.

To most of us, famine connotes a lack of food, leading to starvation. Most familiar, perhaps, are famines the Sahelian region of West Africa or in Bangladesh. However, there is a more insidious kind of famine: the virtually continuous lack of food elements that sustain optimum growth and well-being, that protect against dietary deficiencies, and that strengthen the body against infection. Thus it is useful to distinguish two kinds of food shortage, perennial hunger and catastrophic famine (Mayer, 1976):

Hunger—insufficient caloric, protein, or protective elements on a sustained basis, indefinite in time and often dispersed in space;

Famine—food deficiency, leading to starvation, with a short-term proximate cause, definite duration, and specific spatial locus.

The costs of famine to human society are obvious in the catastrophic case. But we must remember, the high price paid in infant mortality, protein malnutrition, impaired physical and mental well-being, and early death from disease that occur in areas of perennial hunger.

Because food is necessary to survival, it is little wonder that its production, availability, character, and consumption pervade human society. For traditional societies, provision of food is a significant activity and a primary concern for all members. In industrialized societies, increases in inflation inevitably bring the television commentator to the grocery store. Food is sustenance. It is pleasure. Food is the product of the backyard garden; it is a multimillion dollar business. Food is life, politics, matter, energy. Think, for a moment, of the nature of food. Your first thoughts may focus on its nutritional components—calories, protein, vitamins, minerals—recalling your modicum of knowledge of these elements. In addition, you may reflect on the events that are the context of food consumption: the wedding reception, the bowling banquet, Sunday dinner, the quick lunch along the highway, the reverent taking of bread and wine. These events suggested a wider function of food, that is, food and its ingestion as a social act, laden with symbolic value, tying people together. Food symbolizes time of day (bacon and eggs; tea and crumpets), season of year (eggnog and fruitcake; hamburgers and potato salad), and ceremonial events (Thanksgiving turkey). Hot dogs would be as out of place at a White House state dinner as ham at a Bar Mitzvah. What we don't eat is often as meaningful as what we do.

In considering the actual physical characteristics of food, another set of food related functions may be obvious. Color, texture, arrangement on serving dishes, sequence of courses, and other attributes suggest an aesthetic role of food beyond its nutritional role. What about food that has been highly modified in form, such as alcoholic beverages? Is the function of imbibing only nutrition, or is it even nutrition? Does the kind of beverage have symbolic connotations?

Sustaining the pervasive role of food consumption in society are food supply systems, ultimately rooted in the soil and sun, extending from food producers to consumers. In the case of traditional agriculture in developing areas, even the presence of cash crops does not alter an enduring pattern of the farm family supplying most of its own food, producing little surplus as a reliable food source for others. The producing unit is also the consuming unit, at least until

[1] Triage refers to the separation of wounded or injured people into three groups: those who will survive without treatment, those who will survive if treated, and those for whom treatment cannot be successful and who thus are allowed to die. In the context of this paper, triage refers to abandoning nations whose future is hopeless to their own Malthusian fate.

famine strikes and food must be supplied from alternative local, national, or international sources. At an opposite extreme is the farm in the industrialized world, no longer appropriate for "country bumpkin" jokes, with each employee feeding 25 people who live or work elsewhere. Here, food consumption is at the far end of a long and complex chain linking farmer to consumer. Among links in the chain are flows of materials and ideas to the farm from agribusiness and universities; flows of money, inducements, and subsidies to the farm from grain dealers, food processors, and government, with a return flow of farm products; and links between dealers, food processors, wholesalers, and retailers to the ultimate consumer, depending on heavily used transportation networks. Then, between these extremes are food supply systems based on the so-called "green revolution" farm, beneficiary of improved crop varieties and agricultural techniques developed in local and international research institutes. Here local consumption may also persist, but increasing in importance are sources from which farm inputs are secured (seed, fertilizer, chemicals) and markets for disposing of farm production. Cash and material flows become important, as does the institutional framework supporting them. At various places on emerging chains anchored in the green revolution farm are multinational corporations, philanthropic organizations, international agencies, marketing boards, and consumers.

The intent of this paper is to enhance our understanding of the world food problem. For the moment, let us simply define this problem as the apparent inability of the world's people to feed themselves adequately and consistently. The problem is far more complex than that, of course, and its causes (real or imagined) are incredibly intractable. As geographers, we have the advantage of a holistic perspective from which to undertake our analysis—we can see the proverbial forest as well as the trees. Because the world food problem has many forests and innumerable trees, we must focus our discussion on some basic issues that may help us to generate insights as well as place a multiplicity of data into a comprehendible framework. Our specific purposes and approaches are these:

First, we ask whether there is a world food problem, suggesting an affirmative answer;
Second, we explore some dimensions of the world food situation, including a geographical perspective on the factors that contribute to the problem;
Third, we analyze some important aspects of food supply systems, with particular attention to a critical questioning of industrialized agriculture as a solution to the world food problem;
Fourth, we discuss potential solutions to the world food problem, suggesting the risk of increased duality and triage; and
Finally, we conclude that although no panacea exists, there is hope that equitable, humane solutions will prevail.

One of us was once visiting Everglades National Park in Florida. There, a bright and seemingly enthusiastic young park ranger explained the ecology of the area. Subsequently she soberly explained that the life-sustaining water supply to that vast area was threatened, and that here was an excellent opportunity to monitor the impact of human activity compounded by nature's variability. With amazement we noted the ranger's virtual lack of passion as she described past and present ecological devastation caused by man. Later we realized that scientific aloofness and at least superficial indifference make it possible to cope with truths that may be emotionally intolerable. So too, scientific detachment makes discussion of the world food problem possible. However, two forces mitigate against a lack of involvement. First, we are participant observers in the scene. Whether we experience higher food prices or the sense-dulling spectre of dead and dying children photographed in famine-stricken areas, we shall find it increasingly difficult to hide our personal and humanitarian concerns. As students, we examine the problem; as Americans, we *are*, from some viewpoints, the problem. We are reminded of the 1960's slogan, "If you're not part of the solution, you're part of the problem." Second, as geographers we are among those whose intellectual bent is toward human organization of space and use of resources. The current situation calls forth every iota of our personal and professional competence to understand, if not solve, the world food problem.

The spectre of famine is increasingly our daily companion in the world . . . Almost everyone has a different answer (John Deere advertisement, 1976).

I. THE WORLD FOOD PROBLEM

The world food situation is serious, even precarious. It is also true that the world may have, for the first time in history, the ability to deal effectively with the interacting problems of food production, rapid population growth and poverty.

Sterling Wortman (1976:31)

Is There A Problem?

Shortly after its creation, the Food and Agriculture Organization of the United Nations (FAO) issued the first of its *World Food Surveys* (FAO, 1946). Based on an assumed daily caloric requirement of 2600 calories per person, it suggested that two-thirds of the world's population was malnourished. The *Second World Food Survey* (FAO, 1952) took into account regional differences in body size and age-sex distribution of the population, still suggesting that a majority of the world's population went hungry, a view still held by the FAO in 1961 (FAO, 1961) and in its *Third World Food Survey* (FAO, 1963). Pessimism was the theme through the 1950's and 1960's (PSAC, 1967), with famine predicted by 1975. A series of popular works echoed the theme, including books by Vogt (1948), de Castro (1952), Russell (1954), Paddock and Paddock (1964, 1967), Ehrlich (1968), and Dumont and Rosier (1969). By the late 1960's, pessimism had become alarmism in some quarters, whereas official views turned in the opposite direction. Optimism expressed by Bennett (1954), Clark (1967), and Cochrane (1969) was paralleled by official views (Boerma, 1970). In its *State of Food and Agriculture* for 1969, the FAO predicted food surpluses rather than deficits, reflecting a combination of peak world grain stocks accompanied by significant downward revision of human caloric and protein needs by World Health Organization (WHO) and FAO expert groups. The promise of the "green revolution" contributed to a general euphoria (Walters, 1975). In 1970, the U.S. Department of Agriculture's (USDA) Economic Research Service projected a supply of food and fiber in excess of demand through 1980 (USDA, 1970). A major problem seen in the early 1970's was how to dispose of agricultural surpluses (Poleman, 1975). In 1971, an FAO nutrition group again cut caloric needs and protein needs by one-third (FAO, 1971b).

By the middle 1970's, pessimism again emerged. Grain stocks had dwindled as a result of production shortfalls, coincidentally following a substantial decrease in Canadian and American wheat acreage in response to projected surpluses. Ominous predictions about changes in global climate appeared (Bryson, 1976), suggesting that crop failures in northern areas and famines in tropical, semiarid areas could become the norm. Brown (1974) critically assessed the world food situation, and Eckholm (1976) more recently documented ecological decline in major food producing areas. A continuing debate over population growth with respect to resources was encapsulated in computer simulation models of world futures that suggested population growth and economic collapse (Meadows *et al.*, 1972, 1974). Had the FAO overstated food problems in the 1940's and 1950's? Do current FAO food estimates understate supplies by ten to 15 percent, as suggested by Poleman (1975:511)? Was the optimism of the late 1960's unwarranted? Is the present problem overstated (Poleman, 1975)? Is the apparent food problem simply a result of transient factors (production shortfalls and lack of grain reserves) or symptomatic of a more enduring dilemma (Sanderson, 1975; Walters, 1975)? Is the world presently overpopulated (Whittaker and Likens, 1975); about to become overpopulated (Meadows *et al.*, 1972); or capable of supporting double, if not ten times its present numbers (Clark, 1967; deWit, 1967)?

That malnutrition exists is not disputed, nor is the occurrence of famine. What is disputed is whether these events simply reflect an inability of local food systems to overcome social, economic and spatial barriers to equitable production and provision of food, or whether these patterns are harbingers of a future in which hunger and famine will be increasingly commonplace. One result of this disagreement and uncertainty is a plethora of ideas about causes which contribute to the world food problem. It is possible to see a sample of these causes in a geographical framework (Table 1).

At a world scale, a number of pervasive factors creating food problems are suggested. Two environmental perspectives can be discerned. The first argues that food problems, even famine, emerge from occasional vagaries of environment, including

TABLE 1. A SPATIAL TYPOLOGY OF "CAUSES" OF THE WORLD FOOD PROBLEM

Worldwide Problems:
- Occasional natural catastrophes
- Deteriorating world environment
- Imbalances between commodity supply and demand
- Lack of commodity reserves
- Warfare and civil disturbance
- Culturally-based food prejudices
- Declining ecological conditions in food-producing regions

The Industrialized World:
- Excessive use of material and energy resources
- Pollution
- Inefficient, animal protein-based diets
- Insufficient application of science and technology
- Excessive government control
- Insufficient research funds
- Conversion of farmland to competing uses

The Developing World:
- Excessive population growth
- Imbalance among population, resources, technology
- Lack of economic incentives
- Lack of development
- Insufficient government attention to the rural sector
- Inability of traditional food supply systems to cope with change

Industrial-Developing World Links:
- Inequities in access to resources
- Insufficient research and technology transfer
- Lack of development planning
- Insufficient food aid
- Excessive food aid
- Politics of food aid
- Inappropriate technological research
- Lack of institutional structure
- Inappropriate role of multinational corporations
- Insufficient development of agribusiness

Source: Compiled from sources listed in the bibliography.

drought, floods, earthquakes, and other natural disasters. The second argument suggests that these factors may become increasingly probable, that world environmental conditions are deteriorating, particularly in response to human use of resources and anthropogenic pollution.[2] Imbalances between supply and demand on the global scale, accompanied by lack of reserves to meet the needs of catastrophic famine, are also important. So too, patterns of warfare and civil disturbance are worldwide in distribution. Culturally based food prejudices limit use of potentially valuable food resources, or create demands for higher "quality" foods, much less efficiently produced. Finally, there is a widespread decline in the ecological conditions in food producing regions, unfortunately corresponding to ever-increasing demands for productivity.

Some analysts find fault in the system of industrialized agriculture that characterizes much of the Western world. Some argue negatively that the problem *is* industrialized agriculture—that it uses resources profligately, imposes uncompensated pollution costs on the environment, focuses on inefficient food conversion through animals, and diagnostically represents the overabundance, indeed the gluttony, of the industrialized world. Some argue positively, that the world food problem could be solved more rapidly if impediments to the already incredible productivity of industrialized agriculture were removed. Among such impediments are insufficient research and development of technology, conversion of farmland to nonagricultural uses, and excessive government control. As bumper stickers in the Middle West suggest, "If you complain about farmers, don't talk with your mouth full."

Other viewpoints lay blame on the developing world, with a primary focus on excessive population growth, particularly in relation to local resources and farm technology. Traditional farmer perversity, lack of economic incentives, inattention to the rural sector, and lack of development in general are said to contribute to an increasing inability to feed burgeoning populations.

A final perspective addresses the links between the industrial and the developing world. On the one hand, it is argued that food problems are symptomatic of greater inequities in access to resources; that there is insufficient research and technology transfer from the industrial world, a lack of development planning, and insufficient food aid. In contrast, it has also been argued that there has been too much food aid; that technological research has been largely inappropriate to the developing world; and that the population problem is really a problem of excessive resource demands by the industrialized few.

Later we will be looking at a variety of possible solutions to the world food and famine problem. At that time, we shall see that these can be classified in three groups: advancement of technology; population control; and restructuring the world socioeconomic order. Within each group there are a

[2] A large number of observers have suggested that climatic change may occur as a result of pollution by particulates and carbon dioxide. Warming of the earth's atmosphere since the 1880's may be attributed to increased carbon dioxide, which enhances the greenhouse effect. Since 1940, global cooling may be related to particulates, whose reflection of solar radiation may have offset the CO_2 effect. Recent observations in the Southern Hemisphere give credence to these theories. That area appears to be experiencing a contemporary warming trend, which could be related to rapid CO_2 diffusion in the atmosphere; but particulate concentration in the industrialized Northern Hemisphere, which has lead to northern hemispheric cooling, could be mistaken for a worldwide trend (Damon and Kunen, 1976)

number of specific solutions, and both category and specific remedy are predicated by particular notions of the nature of the problem. Those who see the problem as one of excessive government control are unlikely to see government- or internationally-organized redistributing of world wealth as a solution, any more than those who define the problem as inappropriate technological research for developing countries are likely to call for a greater role for the multinational agribusiness corporation. Although there are no clear dichotomies between problem definitions and proposed solutions, the very complexity of the problem suggests a probable conflict of ideas among perspectives.

The World Food Situation

The world food situation in the mid-1970's can be characterized in a few poignant phrases:

(1) Fifteen percent of the world's people malnourished
(2) Declining world food stocks, both in kind and in potential production from idle cropland in major exporting nations
(3) Increasing food and farm input prices
(4) Increased food demand to meet needs of population growth
(5) Further growth of demand due to per capita increases in consumption as income rises
(6) Crop production increasing slightly more than population as a result of modest increases in yields and increases in cultivated land
(7) Declining world grain yields
(8) Increasing world food trade focus on selected industrialized nations
(9) Continuing importance of food aid from the industrialized nations
(10) Significant deviation of average national diets from normal requirements, including both overfed and underfed populations.

To some, these observations suggest our present proximity to the levels of population beyond which the world's population can no longer be fed. To others, they connote a decreased ability for successful national or international response to threat of catastrophic famine. There are few who view the situation with optimism.

Since many of the observations listed are dealt with in materials easily secured elsewhere (e.g. Brown, 1974; USDA, 1974a), we will summarize each briefly before turning to a geographic perspective on the present world food situation. As we shall see, it is extremely difficult to judge nutritional status from national level statistics on population and food production. Data at that level obscure important dietary variations in different locales, among different socioeconomic groups, and among various age-sex groups in the population. Nevertheless, national level food consumption statistics developed by the FAO in the 1960's (FAO, 1971a) and for the industrialized nations by the Organization for Economic Cooperation and Development (OECD, 1973) reflect a conclusion arrived at by many localized dietary surveys: perhaps one out of six of the world's people is malnourished (lacking sufficient caloric or protein supplies), concentrated principally in the developing regions (Table 2). The FAO suggests that one-half of the children in developing countries may be malnourished (USDA, 1974a:50). However, if, (a very big *IF*), only a small proportion of the world's grain production now fed to livestock were available, when and where needed, aggregate dietary insufficiency could disappear. The needed 25 million metric tons of cereals (USDA, 1974a:51) represents less than one-quarter of the grains presently fed to livestock in the United States (OECD, 1973), and an even smaller proportion of the total land used to produce feed grains and forages for livestock that could produce human food.

World food stocks have declined drastically in recent years. Expressed as the number of days' supply of world grain consumption, grain stocks dropped from over a 100-day supply in 1960 to about a 31-day supply in 1976 (Table 3; Brown, 1975:1054). Through the 1960's, actual grain stocks could be increased by one-third to one-half by bringing idle cropland in the United States into production, land

TABLE 2. ESTIMATED MAGNITUDE OF WORLD MALNUTRITION, 1970

Region	Population (billions)	Insufficient Protein/Energy Supply: Percent	Insufficient Protein/Energy Supply: Numbers (millions)
Developed Areas	1.07	3	28
Developing Areas[a]	1.75	25	434
Latin America	0.28	13	36
Far East	1.02	30	301
Near East	0.17	18	30
Africa	0.28	25	67
World Total[a]	2.83	16	462

[a] Excluding Asian centrally planned economies.
Source: USDA (1974a:50), from FAO statistics, 1974.

being held under the Soil Bank, Cropland Adjustment, and other programs (Rasmussen *et al.*, 1976). By 1974, no longer was a seventh of America's cropland held back from grain production by such programs, removing an important "bank account" of potential food production.

Americans spend, on the average, a lower proportion of income on food than residents of other industrialized nations, but food price increases seem as important to us as to others. There are differing opinions about how much these increases really reflect off-the-farm commodity prices rather than price increments in processing and marketing. Nevertheless, recent production shortages, world demand, and high input prices (for fuel and fertilizer, for example) have meant a sharp increase in food prices since 1972. From 1950 to 1970 food prices had been fairly stable, but between 1972 and 1974 wheat and rice prices tripled, and soybean prices more than doubled (Brown, 1974:62). Particularly important are farm input price rises both in industrialized areas and, most critically, in developing areas with already marginal food supplies. From 1971 to 1974 fertilizer prices increased seven or eight times because of higher energy prices and shortages. By early 1976, however, prices had dropped to 1973 levels, only two to four times the 1971 prices, as production capacity met demand. Higher input prices may offset any incentive offered by higher crop prices and add additional foreign exchange burdens to nations with insufficient domestic supplies.

Food demand is projected to increase to nearly twice the 1970 total by the year 2000, both because of population increases and changes in per capita food requirements. Not only will there be some six or seven billion people to feed in 2000, but should modest increases in per capita level of living occur, demand for basic starchy staples will decrease, and demand for livestock products will increase (Chancellor and Goss, 1976:213). If production of meat, eggs, and milk to meet rising demands is based on allocations of resources to animals that could produce food for man, these demands will represent a four- to ten-fold multiplier of food inputs, because of conversion losses in animals.

Through the middle twentieth century, the world (with the exception of certain countries) has kept food production slightly ahead of population growth (Table 4). The developed countries accomplished this primarily by yield expansion, and the developing nations by a combination of yield and area increases (USDA, 1974a). From the base years 1961-1965, the developed countries' growth in population (10.2 percent) was more than offset by expanded production (33 percent). The developing areas had almost as great an expansion of production during the same period (32 percent), but greater population growth kept per capita food increases at a marginal level (Table 4).

Grains supply over one-half of the total world food energy needs. The fairly steady increase in grain yields experienced through the 1960's reversed in the early 1970's (Figure 1). A number of factors may have contributed to this, including weather conditions, release of idle cropland of less than average fertility, higher energy prices, high prices and shortages of fertilizers, shortened fallow cycles, and use of animal wastes for fuel rather than for fertilizer (Brown, 1975:1058). The leveling of agricultural productivity, particularly in industrialized areas, raises questions as to whether yields can be increased further.

World agricultural trade has increasingly focused on North America and Oceania as sources, with Western Europe a continuing "sink;" Latin America and Africa changing from net exporters to net importers of grain; and Asia changing from a marginal net exporter to a major importer (Brown, 1975:1055). Recent increases in wheat production in the major exporting areas have come largely from expansion of land in production. Potential increases in U. S. land in crop production through the end of the century are considered minor, with gains in forest- and pasture-to-cropland conversion and potential new irrigated land offset by nonagricultural competition for land (USDA, 1974b; Zeimetz *et al.*, 1976). The

TABLE 3. WORLD FOOD RESERVES, 1961-1975

Year	Reserves (10^6 metric tons)[a] Grain	Grain Equivalent of Idle U.S. Cropland	Total	Reserves as Days of World Grain Consumption
1961	163	60	231	105
1965	147	71	218	91
1970	188	71	259	89
1975	111	0	111	35

[a] Carry-over stocks at beginning of crop year.
Source: Brown (1975), in *Science*, Vol. 190, December 12, 1975, p. 1054.

TABLE 4. INDICES OF POPULATION AND FOOD PRODUCTION[a]

Calendar Year	Developed Countries			Developing Countries[b]			World[b]		
		Food Production			Food Production			Food Production	
	Population	Total	Per Capita	Population	Total	Per Capita	Population	Total	Per Capita
1955	90.3	81	90	82.5	78	95	85.7	80	93
1960	96.3	96	100	92.8	92	99	94.2	94	100
1965	102.3	104	102	105.0	104	99	103.9	104	100
1970	107.3	119	111	119.0	126	106	114.2	121	106
1973	110.2	133	121	128.5	132	103	120.9	133	110

[a] 1961-1965 levels set at 100. Values for other years may be read as percentages of these base year values.
[b] Excluding Asian centrally planned economies.
Source: USDA (1974a:2).

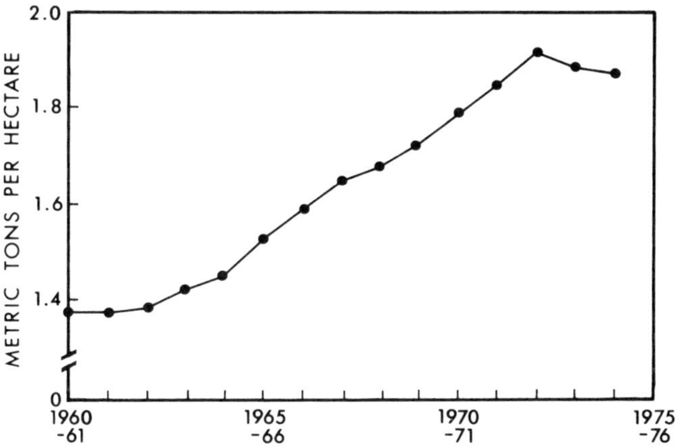

Figure 1. World Grain Yield, 1960-1976. After Brown (1975), in *Science*, Vol. 190, December 12, 1975, p. 1058. Copyright 1975 by the American Association for the Advancement of Science. Used by permission of Lester R. Brown and *Science*.

decades since the middle 1950's have brought an increasing concentration of general world trade in the industrialized nations. In 1955, trade among the developed countries accounted for 45 percent of world trade and 35 percent of world agricultural trade by value. By 1972, the respective figures were 57 and 49 percent (USDA, 1974a:18). This increasing concentration of trade within the developed world is paralleled in the case of grain trade. Trade between developed nations accounted for 40.6 percent of the world grain trade in the late 1950's, and nearly 45 percent by the early 1970's. Today over eighty percent of the world's grain trade originates in the industrialized areas (USDA, 1974a:20).

There are two inescapable conclusions that follow from these data. First, food trade has followed general world trade in an increased concentration within developed nations. Developing nations have experienced a declining trade role which has resulted in a decreased role in international price mechanisms and decreased participation in the flows of world wealth, often accompanied by serious balance-of-payment problems. The old cliché may hold: the rich get richer, the poor get poorer. The second conclusion that emerges from grain flow patterns is the focal role of the developed nations as food suppliers to the centrally planned and developing areas. Concentration of supply is in reality greater; among the developed nations, only a handful are major grain exporters (United States, Canada, Australia, Argentina).

A variety of projections of agricultural demand and production through the middle 1980's (USDA, 1974a) suggests an increasing reliance by the developing countries on importing food from the industrialized areas to make up local production deficits. This food would either be purchased in the world food market or be received as part of food aid packages. The former raises questions of whether the poorest countries have available funds, the latter a wide range of social and political issues, including dependency upon external sources of food and the consequent vulnerability of food supplies to disruption.

Food aid to developing nations includes commodities and financial assistance for rural development, under both bilateral and multilateral arrangements. Among the most important of the bilateral aid flows is the U.S. Agricultural Trade Development and Assistance Act (Public Law 480), passed in 1954. This law gave authority for sale or grant of surplus agricultural commodities for foreign currency, emergency relief, and in exchange for strategic material. Seen in the context of a long series of agricultural legislation since the depression of the 1930's, this law's primary purpose was to adjust domestic agricultural supply to meet demand, and thus to stabilize prices. PL 480 provided for overseas disposal of commodities, a role of increasingly major importance.[3] Relabeled the "Food for Peace Program" in 1966-67, PL 480 from its inception provided some $23 billion in grants and loans (cash value of agricultural commodities) through 1974, of which five billion of the $14.5 billion worth of loans have

[3] Studies by the USDA of the World Food Budget (1961, 1964) all can be seen in retrospect to have exaggerated world food needs; nevertheless hunger was politically desirable in the face of food surplus.

TABLE 5. PERCENTAGE OF U.S. ASSISTANCE RECEIVED BY MAJOR AID AND PL 480 RECIPIENTS[a]

AID (and Predecessor Agencies)				PL 480			
1949-1974		1974		1954-1974		1974	
S. Vietnam	9.5	S. Vietnam	21.2	India	20.2	S. Vietnam	20.7
India	6.6	Cambodia	5.1	Pakistan	7.4	Cambodia	18.8
France	5.6	Indonesia	4.4	S. Korea	7.1	India	7.3
S. Korea	5.3	Pakistan	3.2	S. Vietnam	6.6	Bangladesh	5.2
Pakistan	4.3			Indonesia	3.8	Pakistan	4.4
Turkey	3.7			Brazil	3.6		
				U.A.R.	3.0		
	35.0		33.9		51.7		56.4

[a] Countries receiving 3.0 percent or more of aid category for years specified.
Source: Calculated from data in USAID (1975b).

been repaid. Funds earned by PL 480 sales have been used to pay local U.S. government expenses, to provide loans for development projects, to promote sales of U. S. farm products, and to advance research and education (Cochrane, 1969:134). Because there are various commodities involved (grains, legumes, oils, dairy products), tonnages are an inadequate measure of the volume of PL 480 loans and grants, which reached one-quarter of total U. S. food exports (by value) in the 1960's. By 1974, agricultural exports provided for under government programs comprised less than five percent of total U.S. agricultural exports—$942 million compared to $20,380 million commercial agricultural exports (USDA, 1974a:46).

Title I of PL 480 authorizes sales of agricultural commodities, primarily in local currencies, and represents the largest proportion of the program (two-thirds). Title II provides outright grants of commodities, both directly to governments (one-half of all grants) and through voluntary service organizations (one-half).[4] Although the act was important for surplus disposal, it has nevertheless been a small element in the total U.S. foreign aid picture, never reaching 25 percent of total economic and military assistance, and declining in current years (10.9 percent in 1974).

Funds provided by sales of PL 480 food in recipient countries and other kinds of U.S. foreign aid are aimed at improving local agricultural productivity. The Development Assistance Programs of the U.S. Agency for International Development (AID) include items directly and indirectly related to food and agriculture, including food production and nutrition (54 percent of development assistance expenditures in 1974-75), population planning and health (22 percent) and education and human resources development (11 percent). Approximately one-half of AID budgets are for "security assistance" rather than for "development assistance." Prior to 1969, AID was not allowed to work on food production:

> There was a general belief both in and out of government that other nations should not be encouraged to increase production (of basic food crops) for fear of competition with U.S. efforts to sell its surplus stocks or even give them away (Wortman, 1976:38).

In its presentation in support of the fiscal year 1976 budget, AID argued that bilateral assistance is important for focusing resources:

> —on the key problem areas affecting the poor majority in developing countries in innovative ways;
> —on the countries most seriously affected by the food and energy crises;
> —on problems and areas of critical U.S. foreign policy importance (USAID, 1975a:10).

Although the list of recipient nations and organizations in 1974 (including India, Bangladesh, Sahelian West Africa, Pakistan, Sudan) suggest an important humanitarian role of food and food-related aid from the U.S., the overall flows through AID and PL 480, both in a recent year and during the life of postwar aid, indicate the political motivation of food and food-related aid (Table 5). Brown (1974:69) has correctly observed:

> Although Americans decry the use of petroleum as a political weapon, calling it 'political blackmail,' the United States has been using food aid for political purposes for twenty years—and describing this as 'enlightened diplomacy.'

Food aid, including commodities and grants and loans for economic development, also flows through international organizations as well as through bilateral arrangements. Of the total food aid from developed countries since 1960, the United States has provided eighty percent of the bilateral and multilateral contributions, Canada seven percent, Japan three percent, and France and Germany approximately two percent each (USDA, 1974a:54). European countries are now shifting to multilateral pro-

[4] A list of organizations eligible for this role is prepared annually by the Office of Private and Voluntary Cooperation, Bureau for Population and Humanitarian Assistance, AID (USAID, 1974; *War on Hunger*, July, 1976).

TABLE 6. ANNUAL PER CAPITA GRAIN CONSUMPTION, MIDDLE 1960's

	Consumption[a]			
Nation	Direct	Indirect[b]	Total	Multiple of Tanzania
Canada	202	1791	1993	13.7
United States	200	1441	1641	11.2
U.S.S.R.	344	883	1227	8.4
United Kingdom	169	956	1025	7.0
Argentina	223	625	848	5.8
West Germany	160	588	748	5.1
Mexico	305	242	547	3.8
Japan	320	211	531	3.6
China	312	118	430	3.0
India	288	60	348	2.4
Tanzania	133	13	146	1.0

[a] Pounds per year.
[b] By animal conversion to eggs, milk products, meat.
Sources: FAO (1971a); modified from *By Bread Alone* by Lester R. Brown with Erik P. Eckholm. Copyright © 1974 by The Overseas Development Council. Published by Praeger Publishers, Inc., New York.

grams of the European Community, and considerable moral pressure has been placed on the oil-exporting countries to fund development programs. Major multilateral food programs also include the U.N. and FAO joint World Food Program and the Food Aid Convention of the International Grains Agreement of 1967, the latter serving largely to stabilize world grain trade and provide moral persuasion for food aid to developing countries (USDA, 1974a:54).

A variety of other international organizations provide food-related development aid, including the World Bank Group; Asian, African, and Inter-American Development Banks; the European Economic Community; and programs of the U.N., including the United Nations Development Programs (UNDP), FAO, and WHO. Agriculture has not been a primary focus among all these groups. For example, two of the World Bank Group members, the International Bank for Reconstruction and Development and the International Development Association, jointly allocated only 8.5 percent of their funds to agriculture between 1948 and 1963, 12.3 percent from 1964-1968, and 20.5 percent from 1969 to 1974. During 1975, however, these World Bank donors allocated 31.5 percent of their nearly $6 billion budget to agriculture, with additional funds for population, water supply, and transportation, each of which may assist rural development (World Bank, 1976).

Finally, there is wide divergence in food consumption and dietary sufficiency among the world's peoples (Table 6). The average American consumes 3320 calories per day, about one-quarter more than established needs. Canada's annual per capita grain consumption is nearly one ton, only ten percent of which is consumed directly, the remainder being converted to eggs, milk products, and meat. In the 1960's, Americans and Canadians annually consumed 11 to 14 times as much grain as Tanzanians. Although world food supplies are more than adequate for the world's population, these supplies are neither evenly nor equitably distributed. A geographical assessment of world diets will illustrate this observation.

A Geographical Assessment

No single set of data exists that would allow accurate assessment of the incidence and degree of malnutrition in the world's population. Nutritional needs are understood only partially, and food supply and ultimate consumption are not well documented, particularly for subnational populations. A wide variety of factors that affect individual food needs and consumption are only crudely approximated by national averages and trends (Table 7). Such a multitude of factors affects food needs, demand, and supply that national figures, typically derived from aggregate production and end-of-year carry-over, are potentially misleading, with between-country variation in average diets undoubtedly smaller than within-country extremes. In many developing areas, seasonality of food supply creates marked variation between "hungry seasons" and periods of adequate food supply. Keeping in mind the variation that any statistic may hide, we will assess the broad geographical variations in dietary sufficiency.

Energy and Protein Need

In discussing human food needs, we will focus on energy and protein, two of the most well-known of

TABLE 7. FACTORS AFFECTING NUTRITIONAL STATUS

Need:	
	Age, sex, body size, activity
	Pregnancy and lactation
	Health and biological utilization of food
	Prophylactic or curative needs
	Environment
	Variation around average or typical needs
Demand:	
	Income level
	Income distribution
	Food price
	Government programs
	Education
	Food habits and mores
	Promotion of bottle or breast feeding
Supply:	
	Production of foodstuffs
	Seasonality
	Food processing
	Conversion to animal products
	Marketing efficiency
	Internal and foreign food trade
	Food and aid distribution programs
	Enrichment of nutrient content in foods

Source: Modified from Dwyer and Mayer (1975), in *Science*, Vol. 188, May 9, 1975, p. 567. Copyright 1975 by the American Association for the Advancement of Science. Used by permission of Jean Mayer and *Science*.

our requirements.[5] If sufficient food is available, people will naturally consume as much food energy as required (or more). The FAO and WHO define energy expenditures based upon the needs of an average healthy person in a particular age, sex, or occupational category. It is possible to tabulate the energy needs of a reference or standard man and woman with respect to activity on a unit time basis and to derive idealized daily energy requirements based on a minute-by-minute record of activities (WHO, 1973:109-111). Typically, daily energy needs are also defined on the basis of age, sex, and occupational category. Allowances may be smaller in critical food situations (Table 8). Beyond the body's needs, excess caloric consumption is stored as fat, with about 3500 calories equivalent to one pound (0.45 kg) of excess body weight.

Determination of protein requirements is more difficult. When energy intake is insufficient to meet energy needs, the body uses proteins to provide energy rather than to serve its critical metabolic functions, including growth and tissue replacement. Like energy, protein must be replenished virtually daily. When energy intake is severely restricted, utilization of protein added to diets is impaired:

> When intakes of both energy and protein are grossly inadequate, the provision of protein concentrates or protein-rich food of animal origin may be a costly and inefficient way of improving the diets, since energy can generally be provided more cheaply than protein of good quality. This is an important point in planning programmes for meeting the needs of vulnerable groups in developing countries. Clearly, energy and protein needs should be considered together in planning for the nutritional improvement of populations whose diets are deficient in either (WHO, 1973:19).

Thus energy and protein needs are not independent, and safe levels of protein intake must be specified in reference to energy intake.

The World Health Organization does not specify singular, valid standards for minimum average protein intake. Some attempts have been made to calculate desirable protein/energy ratios. Typical human diets (in absence of hunger or famine) have a surprising stability of 11 to 13 percent of caloric intake as proteins, independent of income or diet composition. Whether this represents the ideal is impossible

[5] Energy needs are essentially quantitative, whereas proteins are a qualitative characteristic of diet, parallel to vitamins, minerals, and fatty acids. Minimum daily allowances have been suggested for many nutrients (Scrimshaw and Young, 1976:62-64). Energy needs are measured in calories, technically termed kilogram-calories or kilocalories.

TABLE 8. CALORIC ALLOWANCES IN FEAST AND FAMINE

Age, Sex, Occupation	Rehabilitation Allowances	Temporary Maintenance	Subsistence
0–2 years	1000–1200	1000	1000
3–5 years	1300–1800	1500	1250
6–9 years	1900–2300	1750	1500
10–17 years	2400–3000	2500	2000
Pregnant and Lactating Women	2200–3000	2500	2000
Sedentary Male	2400–2700	2200	1900
Sedentary Female	2000–2300	1800	1600
Moderate Labor	2500–3000	2500	2000
Heavy Labor	3000–3500	3000	2500
Very Heavy Labor	3500–4000	3500	3000

Source: Modified from Mayer (1975), in *Science*, Vol. 188, May 9, 1975, p. 576. Copyright 1975 by the American Association for the Advancement of Science. Used by permission of Jean Mayer and *Science*.

to suggest. The WHO prefers to specify a safe level of protein intake that will meet physiological needs of nearly all persons in a group, in other words a level of protein *above* average requirements. An additional complication is that protein is not a single substance, but any of a variety of necessary amino acids. The protein content of food can be expressed as "relative protein value" in comparison to human milk or eggs, which have the most complete or highest quality protein based on both amino acid composition and digestibility. Thus, if protein intake is from foods with low relative protein values, one must ingest higher quantities of protein and should eat foods with complementary amino acid composition (Table 9) (Lappé, 1971; Pimentel *et al.*, 1975:754). This does not imply, however, that protein malnutrition follows from a largely or totally vegetarian diet. Complementarity of amino acids from different foods (for example wheat and beans) can assure adequate nutrition (Lappé, 1971). Daily safe protein requirements are based on body weight and other factors (Table 10).

TABLE 9. PROTEIN CONTENT AND UTILIZATION

Food	Percent Protein Content	Percent Net Protein Utilization
Egg	13	94
Milk	4	82
Fish	18–25	80
Cheese	22–36	70
Meat and Poultry	25	68
Grain	8–14	50–70
Rice	8	70
Corn	12	50
Legumes	20–35	40–60
Soybean	35	60
Kidney bean	25	38
Nuts and Seeds	20–30	43–58
Vegetables	2–8	35–60

Source: Assembled from data in Lappé (1971).

TABLE 10. SAFE LEVELS OF PROTEIN INTAKE

Age Group (years)	Typical Body Weight (kg)	Safe Protein Intake (grams protein per person per day)	Adjusted Protein Level Based on Protein Quality with Respect to human milk or eggs[a]	
			Score=80	Score=60
Infants	9.0	14	17	23
1–3	13.4	16	20	27
4–6	20.2	20	26	34
7–9	28.1	25	31	41
Male 10–12	36.9	30	37	50
Male 13–15	51.3	37	46	62
Male 16–19	62.9	38	47	63
Female 10–12	38.0	29	36	48
Female 13–15	49.9	31	39	52
Female 16–19	54.4	30	37	50
Adult Male	65.0	37	46	62
Adult Female	55.0	29	36	48
Pregnant Woman (last 4½ months)		+9	+11	+15
Lactating woman (first 6 months)		+17	+21	+28

[a] Ratio of net utilizable protein to net utilizable protein in human milk or eggs (see Table 9).
Source: WHO (1973:74).

From the levels of energy needs described, it is possible to calculate national food requirements and convert this to average daily values (WHO, 1973). These values have been calculated by the FAO, and average national food supplies can be compared with them.

National Diets

Unfortunately, there are difficult problems in determining national average diet. This is usually undertaken by these calculations for a given year:

+ Food production	(1)
+ Food imports	(2)
+ Carry-over from previous year	(3)
= Food Supply	(1)+(2)+(3)
− Food exports	(4)
− Food fed to animals	(5)
− Carry-over from the current year	(6)
= Disappearance	(1)+(2)+(3)−(4)−(5)−(6)

Disappearance food values are corrected for assumed processing losses, and are then converted to nutritional components expressed as daily per capita values. These calculations are clearly inferior to actual dietary intake studies, which have been undertaken among selected populations in some areas, but they must suffice for a contemporary global assessment.[6]

The FAO made an unusually comprehensive study of annual food balances for the late 1960's (1971a).[7] This information has been partially updated by more recent population and food production data by the FAO and OECD. From these statistics a number of studies can be undertaken on diet composition and adequacy, although analysis based on these data can be amplified usefully by reference to studies of actual dietary intake.

Diet Composition

There is considerable regularity in diet composition as a function of economic development, at least on a national basis (Figure 2). People in poor, developing areas have diets high in starchy staples and low in other components, with regular increases in animal products, fats, and sugars as national income levels increase. In addition, protein supplies change from largely plant to animal sources. Obviously it is not only the developing, largely tropical nations whose diet is composed primarily of starchy staples.

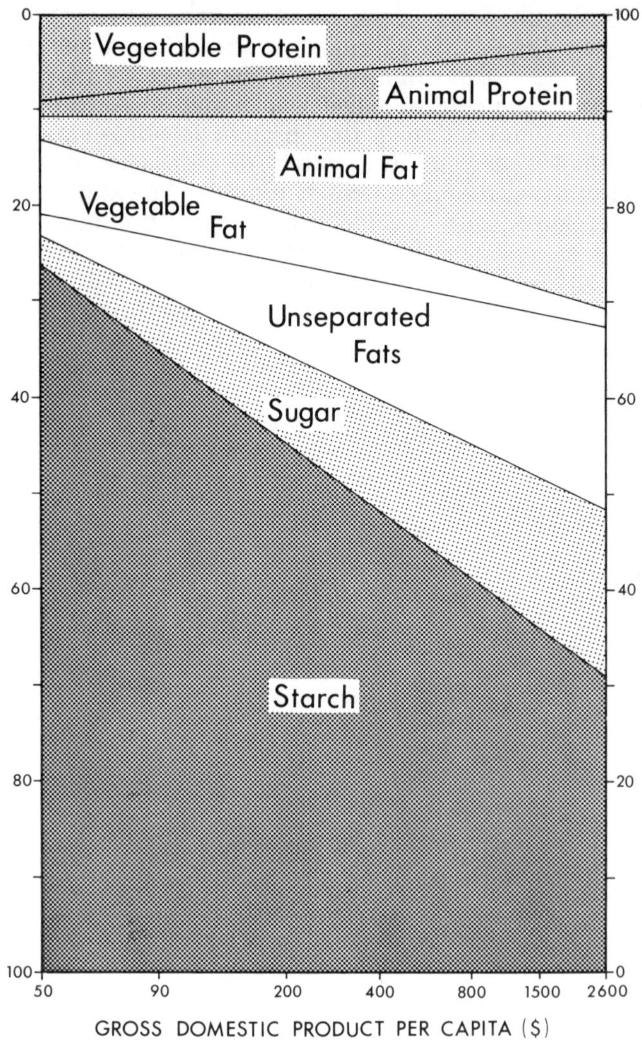

Figure 2. Dietary Composition and National Income, 1962. After Périssé *et al.* (1974).

The Irish farmer at the time of the nineteenth-century potato famine consumed 4.5 kilograms (nearly ten pounds) per day; thus starches made up 82 percent of a typical 3850 daily caloric intake. Protein came from potatoes (45 gm) and other sources (19 gm) (Pimentel *et al.*, 1975:756; Connell, 1950). Similar high starch, low protein diets can be found today among the urban American poor (U.S. Senate, 1974).

A useful means for comparing national diet composition among countries is a graph showing total dietary composition divided among three components—starchy staples, animal-based foods (milk, meat, eggs, animal fats), and other foods—measuring each in calories. For the United States, the average diet is composed of 1255 calories (38 percent) starches, 1326 calories (40 percent) animal sources, and 738 calories (22 percent) from other sources. You will notice that the triangular coordinates add up to 100 percent. To check your understanding, examine the percentages for Tanzania: 68 percent starches, 12 percent livestock sources, and

[6] Among the more comprehensive collections of these surveys are the studies by J.M. May (see May, 1974). The USDA has also undertaken dietary studies in the U.S. with regional and income level differentiation (USDA, 1956, 1969).

[7] Earlier food budget calculations were made by the USDA (1961, 1964).

20 percent other sources (Figure 3).[8]

Percentage values such as those cited are derived from the food composition tables prepared by the FAO and OECD, updated by population and food production indices to derive values applicable to the early 1970's (FAO, 1975b). The triangular graphs illustrate the composition of national diets among the major world regions. The triangle for Africa illustrates the highly starch-dominated diet of that region. Africa's pattern overlaps that of Asia and Latin America. The pattern of European, North American, and Oceanic nations illustrates the importance of animal sources in the diets of the industrialized populations. On each triangle, hollow symbols indicate countries with inadequate diets (those with less than ninety percent of energy needs provided); the relationship of strongly starch-dominated diets in the poorly fed areas is clear. As we noted earlier, this means not only energy deficiency, but probable protein-supply problems as well.

The world map of diet composition (Figure 4) is, in many ways, a map of world wealth and poverty. The map distinguishes four categories of diet composition, from diets with a heavy animal-derived component to those with over 75 percent of energy intake from starchy staples.[9]

Diet Quantity

Lack of uniform, reliable data on diets, accompanied by short-term dietary imbalances, makes any assessment of the status of world diets suspect. Nevertheless, it is possible to sketch a broad picture of quantitative aspects of annual diets at a world scale, so long as we remain aware of the degree to which such an assessment hides variability in diets within countries, ignores problems of diet changes through the year (such as the "hungry season" in West Africa), and relies on the disappearance method of diet estimation.

Basic caloric needs for each of the world's nations, expressed as daily average intake, have been calculated by various world organizations such as the FAO and WHO, using methodologies described earlier. Average diets have been calculated for the late 1960's (FAO, 1971a) and caloric intake for the early 1970's (FAO, 1974) on a worldwide basis; these estimates can be compared with needed caloric supplies. These data may be updated by means of population and agricultural production indices calculated by the FAO and the Foreign Agriculture Service of the U. S. Department of Agriculture (FAO's *Production Yearbook*; USDA's *Agricultural Situation* reports). The world map of dietary sufficiency in the early 1970's (Figure 5) was prepared from these sources in order to reflect, as accurately as possible, the current status of world food consumption on a national basis.[10]

To give some indication of the meaning of values shown on the map, several comparisons are useful. Tanzania's infant mortality rate (deaths before age one per 1000 live births) is, at 139, about ten times that of the United States (Thomas, 1972). In some areas of that country nearly one-quarter of babies born never reach their first birthday. The country has been plagued by food shortages of numerous kinds (Brooke, 1967; Mascarenhas *et al.*, 1973), many requiring international aid. A single half-liter bottle of beer costs a typical Tanzanian a half-day's wages, if he is employed. An equivalent amount of bottled beer in the United States costs about five-minutes of a wage-earner's salary. Each year, Americans consume an equivalent of 160 percent of the annual Tanzanian grain supply in various alcoholic beverages.[11] In 1975 we fed five times Tanzania's total protein needs to cats and dogs.[12] America's dogs, cats, and people are fat, with a price paid not only in excessive food expenditures, but also in ill health.

Food Excess and Deficit

We have alluded to the inefficiency in the consumption of calories and protein from animal rather than plant sources. This observation is based on the dynamics of biological food chains. Think of the flux of food energy as it moves through various species. The solar energy fixed as food by the corn plant is used for the plant's own metabolism and for its growth. The cow that eats the plant uses the major food content for its own metabolism, and a small percentage for its own production of meat or milk. In consuming the meat or milk, much of the food value is again used for our own metabolism, with a desirable percentage for growth (in youth) or an unfortunate

[8] This coordinate system is most commonly used for soil texture types based on sand, silt, and clay composition, but can be used with any classification scheme where three data values add up to 100 percent.

[9] The map demands some study before proceeding. You might answer these questions for yourself:
(a) What kinds of countries eat "high off the hog?"
(b) What kinds of countries have starch dominated diets?
(c) Are there poor countries with an animal-based caloric supply? Why?

[10] Again, the map requires detailed study. Questions like these may guide your analysis:
(a) What countries have the highest caloric intake values compared with needs?
(b) What countries have the lowest values? Middle values?
(c) Are there any modestly, but adequately, fed industrialized nations?
(d) Are there any sufficiently fed developing nations?
(e) What correlation is there, if any, between population numbers and density in developing countries in comparison with caloric adequacy?

[11] Calculated from Brown (1974:39) who cites U.S. consumption at 19.2 pounds (16 kg) of grain per capita per year for beer and liquor. With a base population of approximately 200 million at the time of the FAO food balance data, this represents 3.2 million metric tons per year. For the equivalent base period, Tanzania produced about two million metric tons of grain (FAO, 1971a:517).

[12] Calculated from data in Wittwer, 1975, and Pimentel *et al.*, 1975.

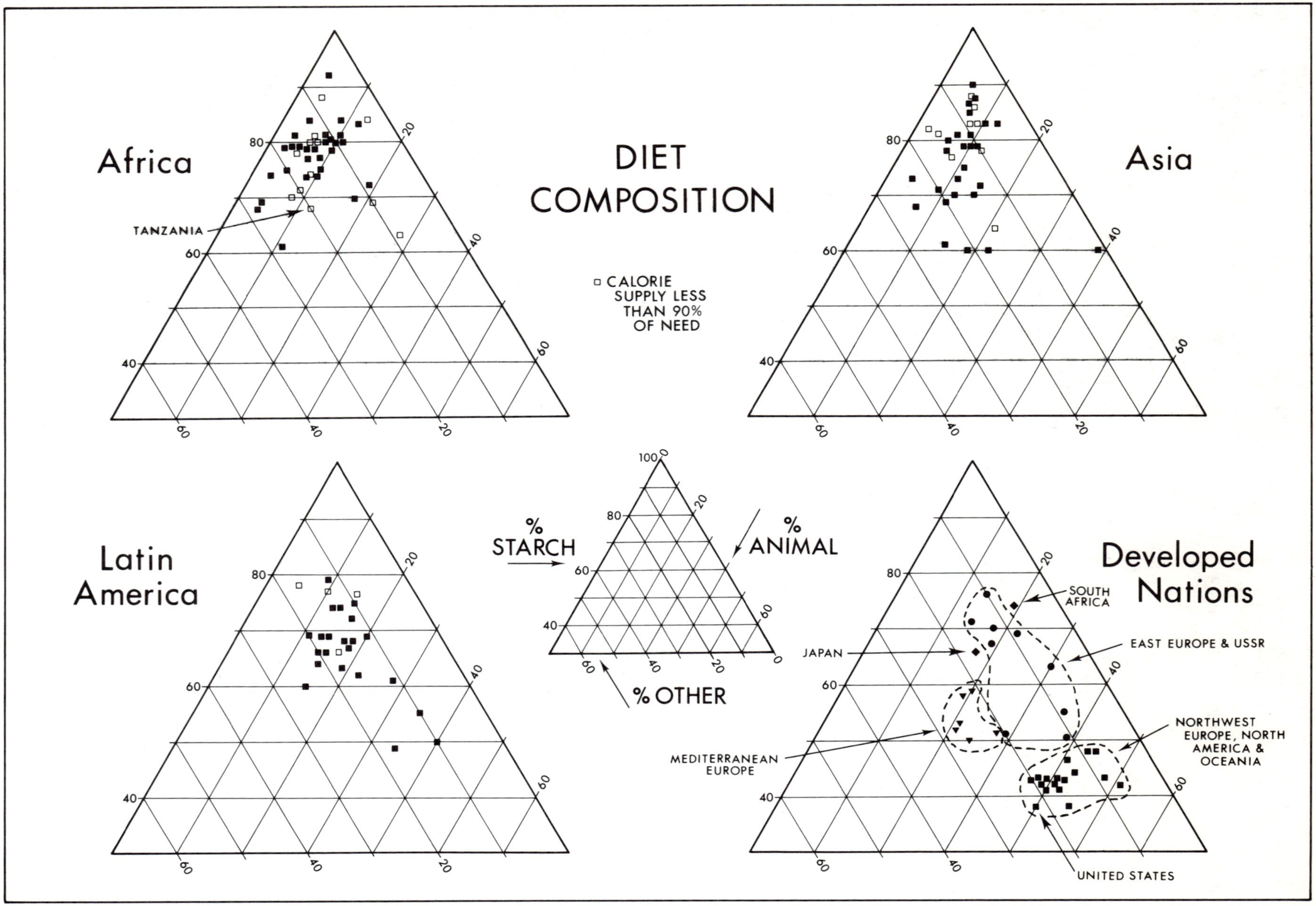

Figure 3. Dietary Composition of National Caloric Intake. Values computed from FAO and OECD data for the early 1970's.

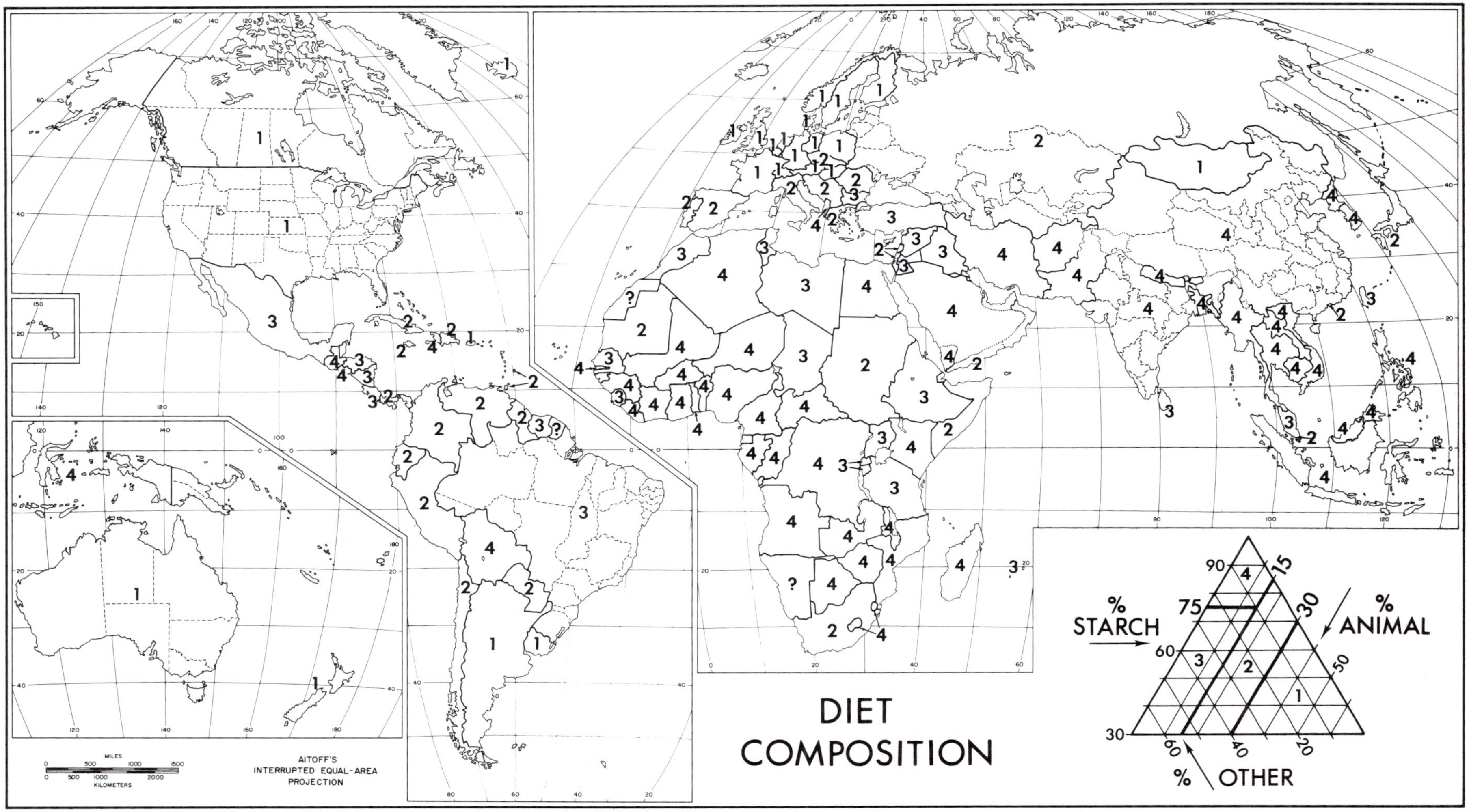

Figure 4. World Pattern of Diet Composition. Coloring the map may enhance recognition of geographic patterns.

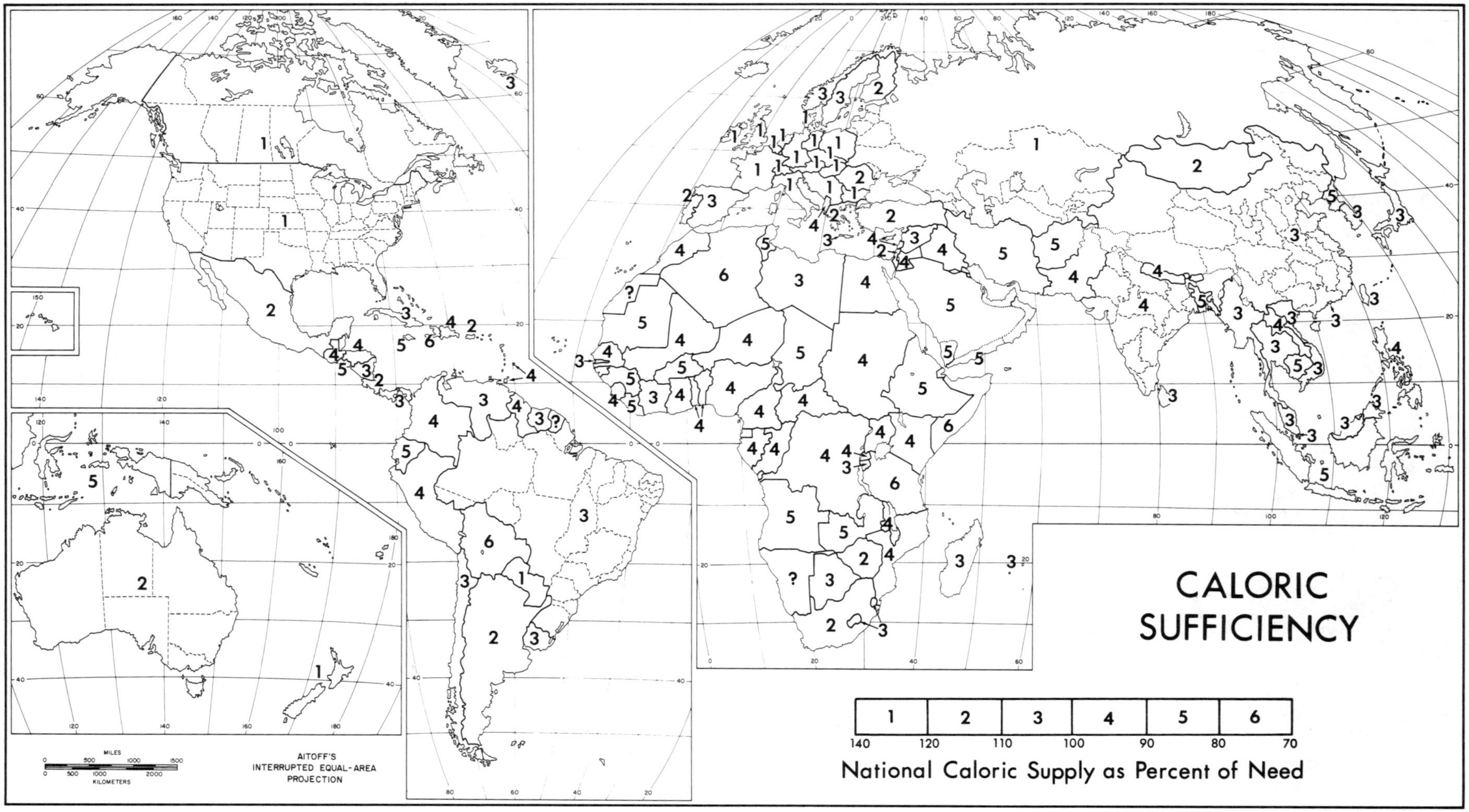

Figure 5. World Pattern of Caloric Sufficiency. Values indicate percentage of national caloric intake compared to needs.

percentage for growth (in adults). At each step along the food chain, food energy is dissipated by normal metabolic processes, as are much of the protein and other qualitative food values. Some plant matter, such as grasses, cannot be used efficiently by humans directly as food, and we are forced to use this kind of material indirectly, through animals. In marginal, semi-arid areas this may be a rational way eventually to harvest solar energy using rainfall levels too sparse for crops, and it may also be rational as part of crop and fallow-pasture rotations in more humid areas. However, when crops produced for livestock could be used for human food, there is a significant loss of total food supply.

This has, of course, been implied in our earlier discussions, but here we must make this "grain or livestock" question explicit. Table 6 shows, for the middle 1960's, the consumption of grain in various parts of the world, both as direct-to-human consumption and also as indirect consumption through animals (with the attendant losses up the food chain). There is considerable controversy over this potential waste of food, which we will take up in our discussion of solutions toward the end of the paper. In assessing the present status of the world food supply, however, we cannot ignore the question. At least one-third of the world's grain production is fed to livestock (Brown, 1974:44).

The same food balances used in the preceding analyses have also been used to calculate the caloric equivalent of food crops fed to livestock in each country. This is a complex accounting task, but it works like this:

+ Calories consumed by people	(1)
+ Caloric value of food crops fed to livestock	(2)
− Caloric value of that food that is passed up the food chain and is already counted in (1)	(3)
= Caloric supply consumed directly and indirectly	(1)+(2)−(3)

Here we are using the concept of indirect calories to mean food energy content fed to animals but lost in the food chain conversion processes. In the United States, over three-quarters of the coarse grains produced (corn, oats, barley) are fed to livestock, as well as wheat, soybeans, and other commodities that could be used directly for human food (setting aside, of course, questions of dietary preference). Values thus calculated for overall caloric consumption per capita may be an understatement, since the human food production potential of land presently producing forage crops has not been included.

The actual values for these calculations are derived from a number of sources, including the food balance sheets (FAO, 1971a); data on crop caloric values (OECD, 1973:xviii-xx); and an assumed ten percent value for conversion efficiency (calories available from livestock per calories fed to livestock). For example, in the United States in 1970, we directly consumed 3319 calories per capita per day. In addition, 136,650,000 metric tons of grains and other human foods were fed to animals. At an approximate 3.6×10^6 calories per metric ton, this is 6685 calories per capita per day. Assuming ten percent conversion, 669 calories were already accounted for in food consumption, so that the total daily figure is 9335 calories, or about 354 percent of average daily requirements! You may already have realized that the approximately twenty percent of total consumed calories (669 of 3319) does not agree with our earlier figure of approximately forty percent animal caloric sources for the American diet. However, half the total feed units consumed by American livestock are from pasture and forage (grass, legume hays, silage). Thus the calculation is reasonably accurate.[13] Furthermore, there is good reason not to calculate the food crop production potential of present forage producing lands. If it becomes important to decrease the direct (fuel) and indirect (fertilizer) uses of energy in industrialized farms, it may still be ecologically desirable to continue the rotational production of legume forage for animals.

For each country, we completed calculations similar to that illustrated for the United States. For many poorer countries, little or no food crops were fed to animals. For others, actual crops fed varied, and conversion factors for caloric values were altered accordingly. For example, many European nations feed high proportions of potatoes, and several Latin American countries feed cassava, plantains, and sugar cane, all with different caloric values which were apportioned with grain values. A world map of dietary balance (Figure 6) shows annual caloric excess or deficit.[14] Direct and indirect caloric consumption extremes include annual deficits of nearly thirty percent and consumption of nearly four-and-one-half times annual caloric needs.

Because of space limitations, and lack of suitable national standards, we have not undertaken similar analyses for protein supply. We suspect that a pro-

[13] A similar calculation for the U.S. was made by Heichel and Frink (1975), indicating an overall food and feed caloric consumption of 17,117 calories per day. This figure differs from our own, since Heichel and Frink included the pasture and forage contribution to livestock as well as meat imports, and calculated back from livestock-based caloric consumption to feed energy as if feed were all from food crops. However, their figure may be a useful indication of the level of potential human food calories lost in producing an American diet, should present pasture and forage land be used to produce crops with caloric yields comparable to forages.

[14] Cook (1976) completed a similar calculation. His approach was similar to that of Heichel and Frink (see Footnote 13), including the total caloric value of livestock conversion efficiencies, without attempting to remove nonhuman foods from the calculation. Here we have assessed only human foods fed to livestock and certainly underestimate the human food potentially lost by forage production on arable land. As with Heichel and Frink, Cook's calculations may be a useful guide to the latter, assuming that forage production provides a utilizable food supply for animals similar to what could be produced on the same land for humans.

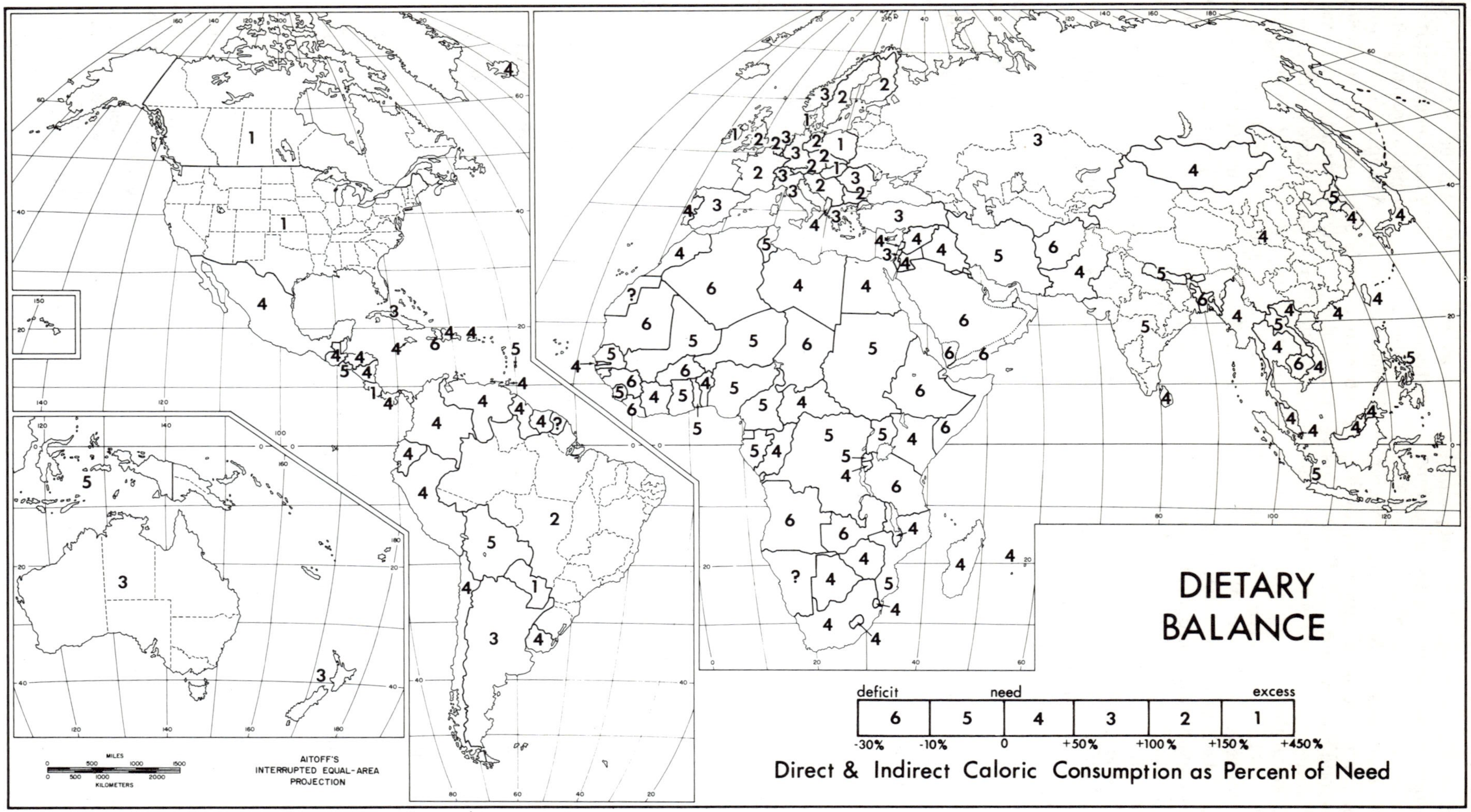

Figure 6. World Pattern of Dietary Balance. Values indicate calories of human food input to the food supply system in comparison to annual caloric needs.

TABLE 11. DIETARY COMPOSITION AND BALANCE[a]

Dietary Balance (percent)	Dietary Composition			
	Over 30% Animal (1)	Over 15% Animal (2)	Under 15% Animal; Under 75% Starch (3)	Under 15% Animal; Over 75% Starch (4)
150 to 450 (1)				
100 to 150 (2)				
50 to 100 (3)				
0 to 50 (4)				
−10 to 0 (5)				
−30 to −10 (6)				

[a] To be compiled by the reader from Figures 4 and 6.

tein map would exhibit similar extremes, with marked deficiency in the poorer countries (in conjunction with the low caloric intakes) and comparable excesses in the industrialized world, particularly if we were to incorporate the protein value of fishmeal fed to animals (and respective conversion losses). American farmers feed about one-half of the annual U. S. fish and fishmeal supply to livestock (Brown, 1974:148).

As a result of this and preceding analyses of the world pattern of food consumption, we might expect to find a significant relationship between dietary composition and dietary balance. The relationship can be demonstrated (Table 11), a task we have left for the reader to complete. For each country, enter data corresponding to its specific dietary composition and caloric excess or deficit. For example, if a country is in the second composition category and second dietary balance category (such as Yugoslavia), it is added to the appropriate section of the table. Completed, the table will allow you to come to some conclusion about world patterns of food availability.

II. FOOD SUPPLY SYSTEMS

Agricultural economists and other agricultural specialists . . . have been probing the world of the developing countries over the past two decades to find the key to successful agricultural development. They need not have traveled so far; the secrets of successful agricultural development are best found in the past history of the U. S.

E. O. Heady (1976:107)

In the Mbeya Region of southwestern Tanzania, a farm family begins preparation for the coming crop season. Although the land is now a parched brown, arrival of the rains will return it to a lush green. From the family's ten hectares (25 acres), they must prepare about three before the rains arrive: cut and burn fallow bush; hoe or plow grassland; recultivate old fields; prune the coffee and spray the plants with pest-inhibiting chemicals. Coffee sales from the last harvest have provided sufficient funds ($225) to pay the children's school fees and to provide for farm and family necessities—chemicals for the coffee, cooking oil, clothing, soap, salt, sugar, and perhaps this year the long-sought radio. The market price of staple foods has nearly doubled since harvest, but food must last until the next harvest so little will be sold. It was once possible to increase productivity by clearing new land, but no new land remains.

As fallows have become shorter, there has been a perceptible decrease in yields, and the family must now prepare larger fields to assure yields with a comfortable margin of safety against drought, pests, or vermin. An ox-drawn plow extends the area that can be cultivated, and weeding food crops and caring for coffee require longer work days than ever before. The family now plants coffee trees on land once available for food crops, but if yields from coffee are sufficient, and high prices hold, the farmer may be able to afford fertilizer to use on food crop fields. However, the price of fertilizer has also increased. Perhaps the area planted to coffee should be expanded. Traditional environmental and agricultural knowledge and the hoe remain the major tools by which the Tanzanian farm family manipulates the environment to produce its food. There is talk of new crop varieties, but the seed must be purchased and the few who have tried them have not done well. There is also talk of resettling people from farms into farm villages, as has taken place in other parts of Tanzania. No longer are people's fates in their own hands and those of family and neighbors. But life is good (Knight, 1974).

Sidamulyo is an incredibily densely populated village in East Java (Prabowo and Sajogyo, 1975). There, a population of over 2000 people must sustain itself on 170 hectares of farmland, one-sixth of which is rented to a local sugar mill each year. The landscape is a finely grained mosaic of agricultural perfection, indicating the intensive labor required to maintain the nearly weed-free fields of rice and other crops. The typical Sidamulyo half-hectare (one-acre) farm must feed a six-member family as well as provide income. Formerly growing traditional and improved local rices, the family now cultivates modern rice varieties (IR8, IR5) from the International Rice Research Institute in Los Banos, Philippines, having followed the leadership of village officials who first planted IR8 in 1968. Since the introduction of modern rice varieties, fertilizer use for local and modern varieties has doubled. The national extension and credit organization known as *Bimas* has encouraged use of insecticides and pesticides, credit for which is provided as vouchers. When former *Bimas* loans are still outstanding, the farmer must turn to outside help, borrowing either in cash or kind at interest rates of up to five percent per month. Because of the higher cost of inputs to modern variety production, more family labor must be used on the field, as less cash is available for hired labor to handle the increased labor requirements of the modern varieties. Backyard production of fruits, vegetables, and livestock provides dietary variety and, with off-farm work, an additional source of income. For the typical Sidamulyo farmer, the arrival of modern varieties has brought marginally lower economic returns, greater reliance on credit, and little change in the occurrence of food shortages. The shorter maturity of the improved rice varieties aids in the intensive double cropping cycle. If the amount of available credit was not tied to use of modern varieties, would they continue to be cultivated?

In rural Pennsylvania, a dairy farmer's major concern is the cost-price squeeze of recent years. Although he has improved crop and milking yields slightly, sharp increases in input costs combined with only modest farm-level commodity price increases mean that he may be working for little return on his own labor, when a reasonable rate of return is assumed on his farm investment. As the Pennsylvania Farmer's Association suggests, a successful farm enterprise must pay all cash expenses, pay interest on equity, provide a return for labor, cover cost of depreciation, compensate for management efforts, and return a profit (PFMB, 1976). By virtually any perspective, the Pennsylvania farm is a big business, with $200,000 worth of land (300 acres or 120 hectares) and capital (buildings, machinery), a hired hand or two, ninety dairy cows, one million pounds (455,000 kg) of annual milk production, an annual cash flow of $100,000, and a net income of $20,000, including a $12,000 assumed return on investment (at six percent interest) and $8,000 for family labor and management. The economics of virtually any part of field operations can be calculated down to the last penny, and agricultural extension personnel monitor farm productivity using computer analyses which guide feed composition and volume, heifer breeding, herd culling, and field operations. Costs and returns can be calculated for any portion of the farm operation. Long, hard hours are a norm for the farmer, and keeping up with latest advisory bulletins and other literature is crucial. Farm radio networks carry daily price developments, and interest is always sparked by latest decisions in Harrisburg and Washington about milk prices and other farm policy. The future of the family farm is a latent issue, with estate taxes, children who prefer an urban life, and possibilities of incorporation for tax purposes as complicating factors. Nevertheless, this farmer and his neighbors, for all their complaining, have a deeply rooted sense of personal value in farming and in feeling a sense of accomplishment in successfully managing so complicated an enterprise.

Fundamental Requisites

These introductory vignettes cover different cultural and technological levels, suggesting a diversity of food production systems. Are there any common features of food supply systems that span the potential range of food supply configurations, from isolated hunting and gathering societies to complex, regionally-specialized industrial agribusiness? One way to answer this question is to suggest a list of basic requirements that any human food supply system must meet, a kind of universal map or chart for structuring our discussion of food systems that are apparently incomparable. Table 12 suggests such a list of fundamental requisites for a food supply system, along with some broadly defined structural elements by which agricultural-based food systems meet these requirements. Here we will distinguish between food production requisites and those related to making food available to consumers. Both *production* and *provision* of food have several requirements, each of which can be described briefly and later elaborated in the context of specific food systems:

(1) *Maintain continuity*: a necessary element in the human food supply system is the genetically- and culturally-encoded information that guides the system;

(2) *Provide space*: solar energy is dispersed over space; thus food production, dependent on photosynthesis, also requires space;

(3) *Manage water*: production depends on green plants, in turn dependent on water; control of where and when production occurs is one means of meeting this requisite, as is irrigation;

(4) *Provide nutrients*: plant and animal growth depend on nutrient supply provided by land management or nutrient supplements;

(5) *Channel solar energy*: both micro- and macrospatial structure, as well as temporal sequencing, guide food production to meet this requirement;

(6) *Control succession*: invasion and competition, normal ecological processes, must be controlled to enhance yields from desirable cultigens;

(7) *Provide protection*: plants and animals must be protected against predators, diseases, and pests;

(8) *Harvest production*: spatially dispersed food production is carried the first step toward consumption; concentration of the usable portion of plant and animal growth or metabolism;

(9) *Transportation*: unless eaten in the field, foods must be carried to the site of consumption, often via many transportation links;

(10) *Storage*: the temporally concentrated production of food is made available to meet the evenly distributed needs for food over time;

(11) *Allocation*: social and economic mechanisms such as markets allocate food to consuming units and individuals;

(12) *Conversion*: food may be converted by animals, fermented, or otherwise altered into more utilizable or desirable forms;

(13) *Preparation*: food is processed into dishes or beverages for consumption, usually near the point of consumption in both time and space;

(14) *Ingestion*: rules of proper food ingestion, or etiquette, control how and when food is consumed.

Food supply systems can be seen to consist of processes that gather solar energy and make it available to humans. Geographically, food production is dispersed in space (an *area* phenomenon), whereas food consumption is clustered at points. In addition, seasonality and variability mean that food production may be concentrated in time, whereas food consumption must be virtually continuous. Thus a food supply system channels solar energy, linking food production and food consumption across time and space.

Seasonality is a fundamental problem for all aspects of food supply since no world environment is without seasonal aspects. Even in the most uniform

TABLE 12. STRUCTURAL ELEMENTS OF A FOOD SUPPLY SYSTEM

Basic Requisites	Structural Elements
Produce Food:	
1. Maintain continuity	Genetic and cultural information
2. Provide space	Spatial organization; land tenure system
3. Manage water	Spatial and temporal structure; irrigation
4. Provide nutrients	Land management or nutrient supplements
5. Channel solar energy	Spatial and temporal structure
6. Control succession	Abatement of invasion and competition
7. Provide protection	Disease and pest control
8. Harvest production	Acquire and concentrate usable productivity
Provide Food:	
9. Transportation	Spatial linkages between production and consumption
10. Storage	Temporal linkages between production and consumption
11. Allocation	Intermediate institutions between production and consumption; markets
12. Conversion	Procedures for making food into assimilable or desirable form
13. Preparation	Processing food into final consumption forms; dietary system
14. Ingestion	Etiquette

of environments there are distinct cycles of plant and animal growth, in response to environmental seasonality that may be undetectable to the naive observer. Variability is similar in importance. Variability may be likened to noise or uncertainty associated with expected environmental and social processes. Vulnerability is the degree to which the requisites of food production and food consumption are subject to disruption beyond the ability of the food supply system to cope with variability.

In this section of our discussion, we have several primary purposes. The first is to suggest how seemingly different food supply systems meet the requirements we have described. The second purpose is to compare (and, on several dimensions, to measure) the differences in *how* these requisites are met in food supply systems characteristic of industrialized, traditional, and developing, "green revolution" societies. Finally, we suggest some of the material, energetic, and informational linkages that constitute a world food system.

The American Food Supply System

No food supply system exists in isolation from a larger social and physical environment, but its ties may be strongest to one or another segment of the larger system. In industrialized agriculture, these ties are strongest to the economy; indeed, farming and food provision in the United States are best viewed as an industry, requiring discussion from both an agricultural and an industrial viewpoint (Figure 7). In the latter sense, the food industry (from farm to consumer) reflects characteristics of the larger economy, with evolutionary structural changes that reflect processes paralleled in other economic domains. The key element in the American industrialized food system is power—power in the literal sense of a reliance on fossil fuels, and power in the figurative sense in the vertical and horizontal structuring of the food system and resultant concentration of economic power.

Food Production and Provision

The American food system is built upon a complex agricultural technology.[15] Food production depends, first, on a system of highly specialized institutions and roles which converge at the farm, including the roles of government, universities, and large corporations. The farmer relies upon a combination of folklore, scientific knowledge, and contact with other factors in the system in making decisions. Among the "other factors" are university extension personnel, sales staff from seed and implement dealers, and information on crop price expectations and the worldwide agricultural situation.[16] The farmer also depends on a network of private and public bodies which maintain genetic stocks. These genetic stocks are often proprietary, with intense competition among seed producers. Both technological and genetic continuity are focused on institutions, which also are foci for new information. New crop varieties may extend the areas suitable for cultivation by having greater tolerance of expected environmental variability. Farmers, as well as consumers, are part of an industrial structure closely controlled by government action. A recent official review of agricultural adjustment policies in the U. S. described the aims of these programs:

[15] For contrasting views of American farm technology, see USDA, 1958; USDA, 1975a; Higbee, 1963; DeMarco and Sechler, 1975; Hightower, 1972; Heady, 1976.

[16] The U.S. government, for example, issues periodic bulletins and reports on agricultural commodity and price status, including *Foreign Agriculture Circulars* on various commodities. Examples of the role of rapid communication include daily crop livestock price reports disseminated by radio networks, such as the Pennsylvania Agri-Broadcasting Network, and by university agricultural extension offices.

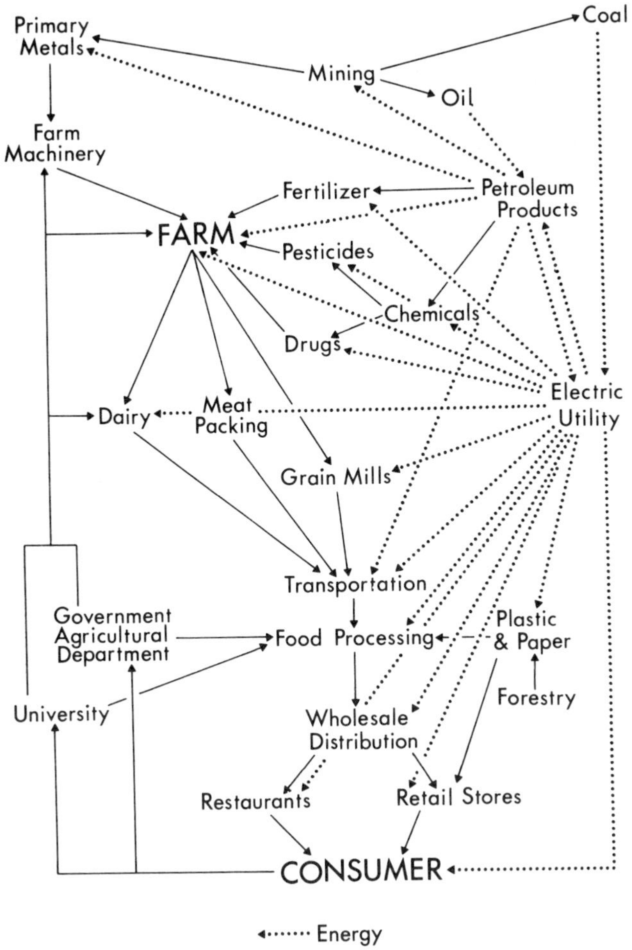

Figure 7. The Structure of the U.S. Food System. Substantially modified from "The Agriculture of the U.S." by Earl O. Heady. Copyright © 1976 by Scientific American, Inc. All rights reserved. Used by permission of W. H. Freeman and Company for *Scientific American*.

> For over 40 years, price support and adjustment programs have had an important impact upon the farm and national economy. Consumers have consistently had a reliable supply of farm products for a smaller proportion of their income than anywhere else in the world. Farmers have been assured of at least specified minimum prices for their products. The legislation and resulting programs have been modified to meet varying conditions of depression, war, and prosperity, and have sought to give farmers, in general, the opportunity to attain economic equality with other segments of the economy (Rasmussen *et al.*, 1976:21).

Just as information is highly institutionalized in the industrialized food supply system, so too is the provision of land for food production. Political allocation in the centrally planned economies and land markets are parallel mechanisms by which space is allocated for production. In the Western world, family farm ownership merges imperceptibly with corporate ownership, since many families incorporate for tax purposes. Many large, vertically structured corporations have substantial farm holdings. Farm size has been an important element in the ability of the farm unit to tolerate economic and environmental variability. These units are made possible by use of energy-subsidized production systems, and bigness can mean sufficient income from good years to provide for poorer years. The small size of farm units during the settling of the Great Plains exacerbated problems of drought (Webb, 1931; Borchert, 1971). Suitcase farming is an adaptation to provision of land under seasonal and variable conditions. This refers to farm ownership in distant locations, with the farm operator traveling among farm units to meet seasonal work demands. This practice is well-known in the Great Plains (Jenks and Kollmorgen, 1958). Suitcase farming extends the land area managed through the year; may increase utilization of expensive machinery by moving it among farms; and provides a spatial solution to variability, since environmental hazards do not occur uniformly, even in the worst drought years of the Great Plains (Thornthwaite, 1941).

For most industrialized farms, water is managed largely by land use practices to control runoff and erosion, still depending on nature's delivery of precipitation. In marginal areas, supplemental or full irrigation may be economically beneficial, and farmers now produce a substantial proportion of such American crops as cotton and vegetables in irrigated areas of the West, frequently with a government subsidy to provide inexpensive irrigation and an energy subsidy for irrigation pumping and spraying. Farmers in subhumid areas practice dryland farming, in which they leave fields in bare soil fallow every other year, in an attempt to store moisture from two years for the benefit of each crop, a productive practice where irrigation is not available, but potentially disastrous when continued drought creates dustbowl conditions from the bare, unprotected soil.

For the industrialized farmer, the soil may be treated largely as a rooting medium for the crop, with the expected nutrient withdrawals provided in the form of chemical fertilizers. Until recent increases in fertilizer prices, it was typically more economical to buy fertilizer than to spread manure, but a combination of environmental regulations on animal wastes and higher fertilizer costs may increase the importance (and benefits) of manure spreading. Chemical fertilizers are another energy subsidy of industrialized farming. It has been estimated, for example, that in California 22 percent of the energy used in farming in 1972 was for fertilizer and that for Canadian grain production, some 46 to 57 percent of the energy inputs are for fertilizer (Chancellor and Goss, 1976). Heichel has estimated that 23 percent of American agricultural use of energy is for fertilizer (Heichel, 1974a). Energy-to-fertilizer conversion factors are available for researchers (Leach, 1975; Commoner *et al.*, 1974; Pimentel *et al.*, 1973), and the cost conscious farmer can check tables of fertilizer costs and expected crop prices to find optimum fertilizer investments (Sommers, 1976:18).

The spatial and temporal structure of the farm must reflect availability of solar energy and channeling of that energy through desirable plants. Although some recent attention has been given to mixed cropping (more than one crop interspersed in a field at one time), industrialized agriculture is largely monocultural, oftentimes not only in the field but also for whole farms and farm regions. Regional specialization and spatial interaction reflect the interaction of crop yields, crop values, and historical tradition (Spencer and Horvath, 1963; Loomis, 1976). The larger spatial system of food production and provision, which pervades many elements of industrialized food supply systems, is a major mechanism for overcoming seasonality and variability problems of food supply. Regional specialization, with transportation and other requisite food provision linkages, provides a supply of perishable commodities over longer time periods, lessening the effects of seasonality. Spatial interaction makes possible bringing food surpluses not only to urban markets but also to areas with food deficits as a result of environmental variability.

Control of succession and protection of crops or livestock in the industrialized system require agricultural chemicals in addition to traditional practices such as plowing or cultivation. Farmers can use chemicals as a variable input in the face of environmental variability. A 1971 sample estimated that over fifty percent of American farms use pesticides, covering eighty percent of corn acreage and 95 percent of rice (Andrilenas, 1975). Machinery is clearly important in meeting the exigencies of farm operations because of seasonality (USDA, 1972a). Custom farm services provide a means for minimizing financial risk because of variability, since each farm operation (planting, spraying, harvesting) is contracted only if the crop appears to be successful. Custom services are also related to seasonality, since custom machinery crews travel in south-to-north orbits "following the sun." An inadvertent means of protecting food supply against damage from hail is the dispersal of production. The farmer may have dispersed production as a result of land bought here, rented there, and share-cropped elsewhere. The food system, wide in spatial extent, similarly spreads the risks.

Harvesting and on-farm crop storage are heavily energy subsidized, including field machinery and natural or liquified petroleum gas used for drying grain. We can look both backward and forward from the harvest. Backward, the harvest is assured, in aggregate, by a number of social subsidies that are an implicit response to variability. These include such direct and obvious subsidies as crop insurance, guaranteed prices, and commodity control programs (Rasmussen, *et al.*, 1976), all of which assure that the participating farmer will not lose his shirt if he ventures to produce. Other subsidies include federally funded irrigation projects, flood control, tax exemptions, assessment variations, land drainage, financial assistance for preparing land for irrigation, the agricultural extension agent network, rural electrification, state aid to schools, housing and welfare for migrant laborers, and government insurance of banks and savings and loan associations.

Looking beyond the harvest, we can see a flow of information constituting a signal from the food provision system to the farmer. That signal is manifest in production contracts, commodity futures markets, and government price predictions, while grain dealers, livestock feeders, milk supply cooperatives, and speculators all encourage production by preharvest purchases. These factors, all of which can be seen as responses to the seasonality and variability imperatives, play an important role in equilibrating prices between harvests and in maintaining, largely in interaction with government, stockpiles over multiple year periods.

The nature of farm inputs reflects recent developments in American farming technology (Table 13). Farm labor has been replaced by machinery, and agricultural chemicals are four times as important now as they were in 1950. Between 1910 and 1970, farm population dropped from over 25 percent to less than five percent; absolute farm production more than doubled; and output per farm worker multiplied nearly eight times (Heady, 1976:110).

That each American farmer produces food for 25 others means a complex system of linkages between the farm and the American dinner table. Food flows over the same transportation networks used by other commodities, and is part of the same industrial and economic system of suppliers, marketplace, and consumers that characterizes the rest of our economy. Many elements of spatial organization and regional specialization characteristic of food production spill over into food provision as solutions to seasonality and variability. In addition, there is opportunity for altering food processing and conversion through livestock as a potential response to extreme disaster, as well as processing to make food available across the seasons. Hoarding, migration, and disaster relief are potential responses to variability, along with the modest although important seasonality of diet to cope with the seasonality of food supply.

Food allocation is largely a matter of market mechanisms, strongly affected by government policy, by various food programs for the poor, and by the oligopolistic structure of the food industry, particularly as seen in the vertical structure made possible by both technological interlinkages and transportation and communication innovations. Given our perspective on the food supply system as an industry, it will be useful to examine one of its major characteristics, that of concentration of ownership.

Although transfers of food are accomplished through economic channels, the basic life supporting role of the commodities involved make it impor-

TABLE 13. INDICES OF FARM INPUTS, 1950-1974[a]

Year	Total	Labor	Real Estate	Mechanical Power and Machinery	Agricultural Chemicals
1950	102	217	104	85	30
1955	103	185	103	98	40
1960	98	145	99	98	50
1965	96	109	99	95	77
1970	101	90	98	100	110
1974	101	83	94	105	138

[a] 1967 = 100 for each column; compare within columns only.
Source: USDA (1975b).

tant to view the food industry through a wider lens. Classical economic discussions are often based on two assumptions: a competitive industry structure within which a large number of similar firms balance each other's influence on production, and prices controlled by consumer influence. Any movement away from competitive structure leads to an erosion of this "consumer democracy." The larger the deviation, the more control shifts from areas external to the industry into the boardroom. For example, when the industry in question consists of a number of small firms and a few much larger firms (oligopolistic structure), the latter groups are in a position to influence prices and also the modes of production characteristic of the industrial sector in question, in ways that are potentially more advantageous to their own profits than to society as a whole:

> Because there are few firms, the actions of one are noticed by the rest; each realizes that any move on its part—a price increase for example—will generate a reaction by other firms. Since the best way to maximize profits is to act as a monopolist would, the oligopolistic firms begin to march to the same corporate drummer (Green *et al.*, 1972:7).

This "conscious parallelism" does not require any formal agreement such as a written contract. Rather, the recognition of similarity of interests in industry results in a close coordination of business policy through independent decisions.

One can examine concentration in the input and processing sectors of the U. S. food system by using U. S. Bureau of the Census (1967) statistics that show the market shares accounted for by the four and eight largest firms in each industrial category (Table 14). Opinions differ as to the level at which concentration leads to price manipulation, with estimates ranging from forty to sixty percent (U. S. Senate, 1974; Green *et al.*, 1972; URPE, 1973). Figures dealing with concentration in input industries may not always show a direct relationship with price, because an input industry may not control all of its own factors of production. Such is the case for the fertilizer industry, which has an apparently low concentration ratio, but depends on supplies of raw materials (particularly natural gas) whose prices are determined by the more highly concentrated energy industry or by the government. Large concentrations are generally associated with overcharging. For example, a study by the Federal Trade Commission reported that:

> If highly concentrated industries were decentralized to the point where the largest four firms control 40 percent or less of an industry's sales, prices would fall by 25 percent or more (Green *et al.*, 1972:14).

The role of input industry concentration is of primary importance to the farmer whereas the structure of processing industries is of interest to the consumer. Each consumer dollar spent on food benefits the farmer but the majority goes to the processor. Thus inflationary trends in food prices do not follow identical trends in input industries. The situation for farmers may be seen in the "cost-price squeeze" from 1950 to the early 1970's. During that period, prices of inputs as well as market prices of food increased substantially, whereas farm prices for raw food commodities fell. The cost of concentration in the agribusiness sector has been calculated by the Federal Trade Commission (Table 15). In 13 food industries, these changes were estimated at over $2 billion in 1972. On the input side, overcharges in farm machinery have been estimated at an annual quarter billion dollars (U.S. Senate, 1974).

Food price increases are sometimes considered to be of little importance because, it is argued, on the average Americans spend a much smaller proportion of their income on food than do people of other nations. For example, average figures show that Americans spend only 15 percent of their income on food, whereas English spend 27 percent, French 31 percent, and Russians 53 percent (URPE, 1973). But average figures are misleading. Since the top five percent of the U. S. population receives twenty percent of all income, whereas the bottom eighty percent receives 55 percent, in reality most families must spend much more than 15 percent of their income on food (Table 16).

Between the large food provision industries and the consumer lies another group of economic institutions, food retailers. Nationally, the top four and top eight firm concentrations of retail grocers were 20.1

TABLE 14. CONCENTRATION IN AGRICULTURAL INPUT AND PROCESSING INDUSTRIES, 1967

Industrial Code (SIC)	Product Class	Percent of Total Sales	
		Top Four firms	Top Eight Firms
Input:			
20421	Poultry feed, including supplements	28	37
20422	Livestock feed, including supplements	28	39
2871	Fertilizers	33	52
28790	Agricultural pesticides and other agricultural chemicals	39	56
3522	Farm machinery	45	56
35221	Wheel tractors and attachments	75	98
35225	Harvesting machinery	67	82
Processing:			
2011	Meat packing plants	27	38
20118	Canned meat (except pet food)	57	82
2015	Poultry dressing plants	15	23
20221	Natural cheese	38	44
20222	Processed cheese	72	47
2023	Condensed and evaporated milk	35	47
2026	Fluid milk and related products	21	29
20321	Canned baby foods	93	98
2033	Canned fruits and vegetables	23	35
20411	Wheat flour, except flour mixes	37	55
20430	Cereal breakfast foods	82	94
20511	Bread and bread type rolls	28	38
20620	Refined cane sugar	59	83
20630	Refined beet sugar	67	96
2072	Chocolate and cocoa products	74	87
20962	Margarine	47	72

Source: U. S. Bureau of the Census (1967).

TABLE 15. MONOPOLY OVERCHARGES

Industry	Market Concentration (1966—Top Four Firms)	Monopoly Overcharge $ Million (1972)
Meat packing	40	$ 483.9
Fluid milk	60	256.7
Soft drinks	90	247.8
Malt liquors	65	198.0
Bread and related products	50	191.9
Canned fruits and vegetables	40	143.6
Confectionary products	40–50	94.4
Flour and other grain mill products	45	88.5
Distilled liquor	55	88.3
Frozen fruits and vegetables	40–50	84.9
Cane sugar refining	40	71.5
Canned specialties	80	71.2
Crackers and cookies	70	57.3

Source: Federal Trade Commission data, quoted by Hightower (1975:64-65). Taken by permission from *Eat Your Heart Out* by Jim Hightower. © 1975 by Jim Hightower. Used by permission of Crown Publishers, Inc.

and 28.2 percent in 1970 (USDA, 1972b:96). National values again are misleading, because local concentration may be much greater. Typical values of metropolitan area dominance by the top four retailers range from 51 percent in Binghamton, New York to 86 percent in Little Rock, Arkansas (Hightower, 1975:21). Immobile, urban poor may be tied to fewer alternatives—more expensive smaller retailers among them—than the affluent suburbanite, compounding the problem of low income and higher proportion of income needed for adequate nutrition.

The inability of the American food supply system to assure adequate dietary quality for 13 million Americans (U.S. Senate, 1973) reflects both the fail-

TABLE 16. AMERICAN EXPENDITURES AS A PERCENTAGE OF SPENDABLE EARNINGS, 1973

Annual Gross Income	Low Cost Plan (percent)	Moderate Plan (percent)	Liberal Plan (percent)
3,328	62	79	97
5,000	42	54	66
7,280	35	45	55
10,000	26	33	41
15,000	18	18	28
25,000	12	14	18

Source: *Christian Science Monitor*, quoted by Hightower (1975:61). Taken by permission from *Eat Your Heart Out* by Jim Hightower. © 1975 by Jim Hightower. Used by permission of Crown Publishers, Inc.

ure of commodity food programs, food stamps, and other welfare programs, as well as the food supply system itself. For the affluent, the food supply system provides commodities virtually on demand, with seasonality and variability reflected in minor price fluctuations that are insignificant with respect to income; for the poor it fails miserably. The Senate Select Committee on Nutrition and Human Needs (U. S. Senate, 1973) has suggested the distribution of hunger in America on a county scale of resolution. Counties with one-quarter or more of the population below poverty income levels and one-third or less of the poor receiving food assistance were classed as "Hunger Counties, 1973." Even counties falling outside these dual measures have significant numbers of poor lacking food assistance (Figure 8).[17] Only one-half of America's poor would appear to be fed adequately.

Having already criticized some aspects of the industrialized food system, it is appropriate to suggest some measures by which its performance might be assessed, and by which other food supply systems might be evaluated. The following yardsticks will be useful for our discussion: land, labor, capital and energy.

[17] An earlier study, *Hunger U.S.A.*, uses a similar approach to suggest patterns of hunger in 1968 (Citizen's Board of Inquiry into Hunger and Malnutrition, 1968).

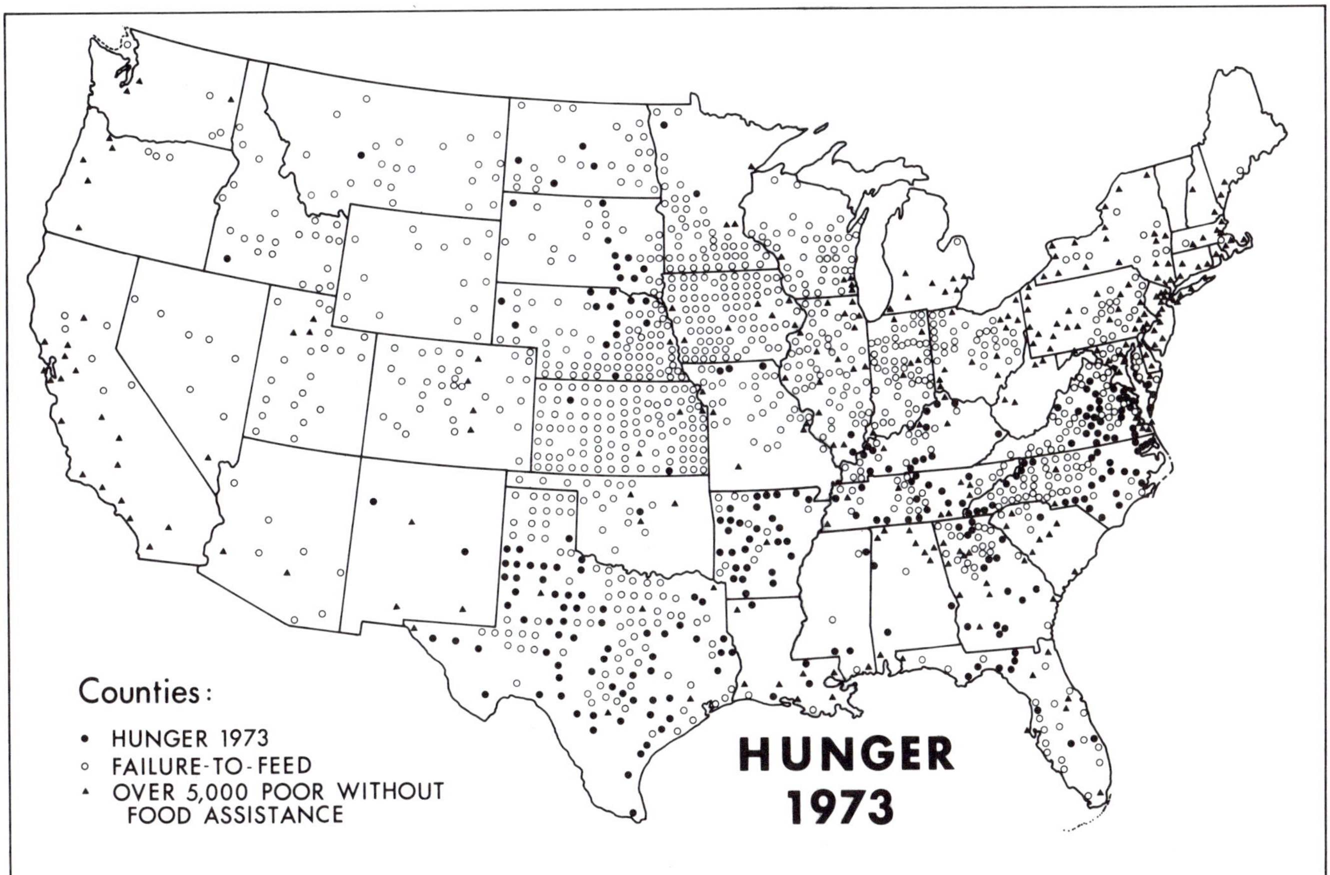

Figure 8. Hunger in the United States, 1973. Failure-to-feed counties have less than one-third of the poor receiving food assistance; hunger counties are failure-to-feed counties in which more than one-quarter of the population is poor. After U.S. Senate, 1973.

Land

Crop yields in the industrialized agricultural system have benefited both from a highly developed technology and from an environment of comparatively high productivity as seen from a world perspective. Although the U. S. agricultural system appears to work miracles (American yields are typically double world standards—Table 17), farmers achieve these high yields by an extremely disproportionate use of agricultural inputs such as fertilizers and farm machinery. Using one-fifth of the world's annual fertilizer consumption and plowing its fields with over one-quarter of the world's tractors, it is little wonder that America's agricultural technology has been so successful in feeding a bit more than five percent of the world's population (Table 18).

The industrialized nations occupy some of the world's most productive agricultural regions, par-

TABLE 17. AVERAGE CROP YIELDS[a] FOR BASIC STAPLE FOODS

Area	Wheat		Rice		Maize (corn)		Roots & Tubers[b]		Potatoes	
	1961-5	1974	1961-5	1974	1961-5	1974	1961-5	1974	1961-5	1974
Developed Market Economies	1.74	2.14	4.91	5.58	3.51	4.01	18.24	22.27	18.37	22.75
North America	1.58	1.75	4.37	4.98	4.17	4.48	20.43	25.62	21.59	26.56
Western Europe	2.17	3.37	5.03	5.53	2.49	4.18	18.00	21.97	18.01	22.00
Oceania	1.25	1.29	6.17	6.10	2.11	3.58	15.57	20.64	15.60	20.76
Other	1.24	1.28	5.01	5.84	1.26	1.79	17.46	17.20	16.18	19.77
Developing Market Economies	0.97	1.17	1.63	1.87	1.14	1.28	7.90	6.91	7.47	8.85
Africa	0.70	0.73	1.28	1.31	0.93	1.11	6.96	5.17	6.15	6.12
Latin America	1.42	1.49	1.73	1.94	1.23	1.39	10.45	11.34	7.19	8.93
Near East	0.99	1.16	3.41	3.62	1.90	2.42	7.30	7.87	10.71	12.23
Far East	0.84	1.18	1.61	1.87	1.00	1.05	7.63	8.21	7.06	8.37
Other	1.67	0.83	1.82	1.96	1.38	2.10	7.38	8.23	7.98	8.10
Centrally Planned Economies	1.01	1.52	2.74	3.19	2.27	2.95	9.76	10.90	10.57	12.04
Asia	0.88	1.27	2.75	3.19	2.41	2.85	8.41	9.52	8.33	9.87
Eastern Europe and USSR	1.06	1.63	2.48	3.70	2.15	3.07	11.13	12.76	11.13	12.76
United States	1.70	1.84	4.37	4.98	4.17	4.48	20.95	26.47	22.42	27.61
World	1.21	1.60	2.05	2.36	2.17	2.51	10.08	9.90	11.93	13.39

[a] Thousands of kg/hectare.
[b] Sweet potatoes, yam, cassava, taro, starchy roots, and tubers other than potatoes.
Source: FAO *Production Yearbook 1974* (FAO, 1975a).

TABLE 18. WORLD CONSUMPTION OF AGRICULTURAL INPUTS, 1973

	Fertilizers									
	Nitrogen		Phosphate		Potash		Tractors		Population (1974)	
Area	Amount*	Percent	Amount*	Percent	Amount*	Percent	Number (million)	Percent	Number (million)	Percent
Developed Market Economies	17.82	46.1	13.96	57.6	11.48	55.5	11.78	71.7	750	19.2
North America	8.77	22.7	5.08	20.9	4.82	23.3	4.99	30.4	235	6.0
Western Europe	7.75	20.0	6.12	25.2	5.56	26.9	5.88	35.8	365	9.4
Oceania	0.21	0.5	1.62	6.7	0.28	1.4	0.44	2.7	16	0.4
Other	1.08	2.8	1.13	4.7	0.82	4.0	0.47	2.9	134	3.4
Developing Market Economies	7.05	18.2	3.45	14.2	1.90	9.2	1.40	8.5	1927	49.3
Africa	0.43	1.1	0.31	1.3	0.19	0.9	0.15	0.9	304	7.8
Latin America	1.80	4.7	1.39	5.7	0.90	4.4	0.72	4.4	317	8.1
Near East	1.21	3.1	0.54	2.2	0.04	0.2	0.25	1.5	193	4.9
Far East	3.60	8.0	1.21	5.0	0.77	3.7	0.27	1.6	1109	28.4
Other	—	—	—	—	—	—	—	—	5	0.1
Centrally Planned Economies	13.79	35.7	6.85	28.2	7.32	35.4	3.25	19.8	1228	31.4
Asia	4.07	10.5	1.56	6.4	0.58	2.8	0.20	1.2	868	22.2
Eastern Europe and USSR	9.71	25.1	5.29	21.8	6.74	32.6	3.05	18.6	360	9.2
United States	8.28	21.4	4.60	19.0	4.61	22.3	4.38	26.7	212	5.4
World	38.66	100.0	24.25	100.0	20.70	100.0	16.42	100.0	3905	100.0

* In million metric tons.
Percentages calculated by the authors; world totals may not agree with sums because of rounding error.
Source: FAO *Production Yearbook 1974* (FAO, 1975a).

ticularly with respect to climatic conditions for growing staple grain crops. Compared with the humid tropics, middle latitude areas may have as much as thirty percent greater photosynthetic potential for the summer four-month growing season (Table 19). This is because potential photosynthesis is a function of the positive effects of day length and solar light intensity (during which photosynthetic products accumulate) and the negative effects of temperature and night length (use of photosynthates for respiration depends on temperature; long, warm nights mean greater respiration with no photosynthesis). This reasoning, documented by Chang (1968; 1970), partially explains the much higher rice yields in the U. S. compared with tropical Asia. Only if two or three crops can be grown per year on a given land area can tropical areas outproduce middle latitude areas.[18]

If industrial technology permits high yields from land, ownership of land resources is an important element in understanding the structure of the American food system. Ownership of land is an economic and political issue. Industrialized farm units are becoming increasingly larger. In 1880 the average farm size was 134 acres. In 1940 it was 175 acres. Since then it has risen at a much faster rate, so that in 1969 it was 389 acres. In 1969, over 54 percent of America's farm acreage was in farms of 1000 acres or more (USDA, 1974b:23). Thus, the recent increase has partly resulted from the emergence of a number of large units. California is perhaps the most extreme case of concentration, partly as a result of its history of large land units dating back to the Spanish colonial system, but the phenomena is sufficiently widespread to cause concern in other regions. In North Dakota, a 1932 law excluded large nonfarm corporation investment in basic agricultural production. California, Florida, and Texas are the major corporate farming states, with half of their 4800 corporate farms having annual sales exceeding $100,000 (USDA, 1974b:26).

Concentration of production in large units parallels their development. Between 1964 and 1970, farms with over $20,000 in sales increased in dominance from 64 to 76 percent of farm sales, although they numbered only 21.5 percent of all farms. Concentration is particularly notable in several farm products. By the middle 1960's, over two-thirds of America's vegetable production was concentrated in farms with over $100,000 in sales, as was one-third of U. S. poultry production (USDA, 1974b:26). Even with farm products where concentration is not as great, such as cash grains, concentration may appear elsewhere along the food supply linkage: six companies control over ninety percent of the wholesale grain market. A 1967 study reported that full economies of scale in many types of American farming could be achieved with farm units of modest size operated by one or two people, with a high degree of mechanization and with custom farm services aiding smaller farms to achieve similar economies (Madden, 1967). In 1969, farms with sales greater than $100,000—0.9 percent of all farms—absorbed 29 percent of all feed inputs, 11 percent of all fuel inputs, 39 percent of all livestock and poultry bought on the market, 24 percent of hired machinery, 17 percent of all seed, 16 percent of all fertilizer, and 41 percent of all hired labor (Hightower, 1973).

Labor

As technological inputs to farming have increased, human labor has drastically decreased (Table 13). The modern industrial farmer in the United Kingdom produces from 125 to 800 times as many calories as he eats by using mechanization and energy subsidies (Leach, 1975). So long as energy is cheap, he can produce food inexpensively and still earn a reasonable wage.

There is a large disparity between farm and nonfarm incomes in the United States ($6,400 versus $9,600 average family income in 1969). These fig-

[18] Potential photosynthesis for various time periods are compared in Table 19 to the Mediterranean (winter rainfall) areas which have the world's highest annual value. Measurements are based on a standard plant, and cannot be compared directly to specific crop yields. For other than relative interpretation of climatic zones, see Chang, 1970.

TABLE 19. POTENTIAL NET PHOTOSYNTHESIS OF WORLD CLIMATIC REGIONS

	Potential Photosynthesis[a]					
	4 Month		8 Month		Annual	
Region (Köppen Climatic Notation)	Amount	% Cs	Amount	% Cs	Amount	% Cs
Humid tropical areas (Af, Am)	2.9	75	5.8	84	8.8	92
Tropical savanna areas (Aw)	3.2	84	6.2	86	9.2	97
Warm, moist subtropics (Cfa, Cwa)	3.4	89	6.7	92	9.1	95
Humid, midlatitude west coasts (Cfb, Cwb)	4.0	105	7.2	100	8.7	91
Mediterranean climates (Cs)	3.9	100	7.2	100	9.6	100
Cool continental climates (Dfa, Dwa)	3.8	99	6.4	89	6.8	71
Cold continental climates (Dfb, Dwb)	4.1	107	5.8	86	6.2	65

[a] Kg/m^2 for stated period. Values attributed to all periods assume adequate moisture availability.
Source: Modified from Chang (1970:96). Used by permission of the *Annals of the Association of American Geographers.*

ures are deceiving because much of the disparity can be attributed to the very low income levels of farm labor. Much of this labor is invisible to urban Americans, since it is expended far away, on crops seldom conjuring images of American farming. Farm laborers pick fruit, clean and bag vegetables in the field, spray pesticides, weed, and do other labor intensive operations. By 1972, only 2.8 million workers remained in the U. S. food production system; approximately twenty percent of these were black, Puerto Rican, Mexican-American, or Mexican. Of this number, the minority who work all year (13.1 percent) made an annual average of $3,170, whereas the overall average wage was only $1,160 per annum. Government policy aids this process of human abuse by differentiating between the rights of agricultural and nonagricultural workers. Large numbers of the former are excluded from many of the benefits of labor legislation. Often, farm workers are not unionized, and migrant workers are not allowed to bargain collectively for improved benefits such as unemployment compensation. In 1972, federal minimum wage regulations covered only 535,000 of these workers (URPE, 1973:70-71,98). Agricultural workers receive low rates of injury compensation, and in thirty states are granted none at all, even though agriculture is the third most hazardous industry after mining and construction. Moreover, they receive less comprehensive social security coverage than workers in other industrial sectors. Also, weak child labor laws allow about 800,000 children to work during school hours (URPE, 1973:70-71).

A justification for this lack of power among workers in the agricultural production sector advanced in Congress is that:

> . . . farm commodity prices are determined on a day to day basis in a highly competitive world market and rigid bargaining legislation might weaken the ability of U. S. agriculture to compete in world trade (U. S. Senate, 1972:50).

Other factors reinforce this mythology. The trend with some crops towards "runaway fields" is one example. To avoid increased labor costs, corporate farmers may move their fruit and vegetable production to southeast Asia or Latin America (DeMarco and Sechler, 1975:80-82).

Our image, then, of the highly productive American farming system is tainted by recognition of the poverty levels of farm laborers. U. S. policy encourages further development of agricultural technology, some intended further to replace the farm worker, rather than to improve conditions for this group. We subsidize the exploitation of farm laborers by public provision of welfare, migrant housing, and other programs which make possible the continued payment of substandard wages.

Capital

The industrialized farm is a capital intensive enterprise, depending on long-term capital availability for land, buildings, and machinery, and short term capital for operating costs between harvests (Evans and Simunek, 1976). Since virtually all segments of the farming system (land, labor, energy) have monetary value, detailed analyses of farm and crop budgets have been prepared to guide farmers in their decisions.[19] Assuming that money could earn a return by being invested in alternative endeavors (such as a bank account, factory, poker game), any investment in agriculture represents an investment that must yield a return at least as great as the opportunity cost of the money, that is, what it would earn elsewhere. This is why we separated the annual income of the Pennsylvania farmer cited earlier into return on investment and payment for labor and management. When capital must come from private sources, reasonable return is expected. It is interesting to note that in the early 1960's, even assuming a very modest (4.1%) return on capital investment, many American farmers were working for wages as low as 50¢ per hour, and some, in areas of variable crop yields, were actually losing as much as one dollar for every hour worked (Higbee, 1963: 167-169).

Many industrialized farms belong to cooperative organizations for supply of inputs or sale of production, and others assure markets by contracting their harvest before production. Some data are available for an analysis of contract farming, the system whereby a corporation contracts for use of a farmer's land and production resources, gaining the cheap labor advantages of family farming, without assuming the risk of ownership or capital investment. The farmer sells his produce to the purchaser or processor at a price fixed in advance between the two parties. Contracts vary greatly in their details and whether the farmer gains or loses as a result depends on environmental conditions during the harvest year. Some of the worst examples of oppressive contracts are found in the broiler poultry industry:

> The grower's contract is so unequal that he has been compared to the sharecropper in both his status and his poverty. The difference in this vision of that infamous relationship is that the bossman is an absentee landlord with a corporate, not a personal, identity, whose "big house" is in a panelled boardroom in far away Minneapolis or New York. Otherwise the analogy needs no explanation. The "cropper" is visited by the "field man" who supplies him with the essentials to make a crop (in this case baby chicks, feed and technical advice); the field man collects the harvest (8 week old chickens), measures the results (he weighs the chickens and sets the price) and decides what the croppers share ought to be. The cropper may even be forced to buy his supplies (his broiler equipment) from the company store. The cor-

[19] The most detailed of these budget calculations are in the Firm Enterprise Data System of the Economic Research Service of the USDA. They have a high degree of spatial resolution and budget detail. Whole farm budgets are also prepared by the USDA, as well as by various farm organizations and agricultural extension services (USDA, 1971, 1976; Krenz *et al.*, 1976).

porate landlord keeps all the records . . . (U.S. Senate Subcommittee on Monopoly, 1972:3702).

Government estimates of the share of U. S. food production produced under contracts indicate that for some commodities, such as milk and broilers, 95 percent of production is controlled by contracts, with considerable variation in other commodities (U.S. Senate, 1972). Such contracts may exert a stabilizing effect on food supplies, but they also suggest a further dimension of concentration, control, and exploitation.

The political importance of food forces the government to support agriculture by policy and by subsidy programs. Schultze (1971) estimated that agricultural subsidies cost the U. S. taxpayer $9 to $10 billion annually. Since the 1950's there has been an overall increase in both the number of subsidy programs and the level of direct payments under them. Direct payments included commodity control programs, land retirement payments, and conservation and soil bank payments for land improvements. In addition there are a number of other subsidies from the government, including the Smith-Lever (1914) and Hatch (1887) Act grants to universities for agricultural research and extension; tax reductions for research and development costs; subsidized irrigation water supplies; and government research and agricultural extension roles. Price supports and direct payments have disproportionately benefited large farms. An estimated 53 percent of price support benefits were received by Class I farms in 1969, as well as 29 percent of direct payments, although these farms, with $40,000 or more in annual sales, constitute only five percent of all farms. Benefits for these large farms has been estimated at over fifty percent of their total income, but less than one-third the income of the farm group with lowest income can be attributed to subsidies (Schultze, 1971). The spatial distribution of government agricultural benefits parallels the distribution of farm size, with the modification that crops which dominate some areas are not subject to government support. In 1970 the ten largest single payments, totalling $18.3 million, went to California ($15 million), Hawaii ($2.3 million), and Florida ($1 million) (*Ramparts*, 1971).

The distribution of subsidies may be compared with government tax receipts from the agricultural sector. Tax revenues obviously need not necessarily remain intrasectorial in distribution, but the comparison is of interest because it reflects another way in which the government has supported the trend toward increasing farm size. Small farms with low levels of income generally have a lower percentage of reported losses for tax purposes than farms with high incomes. The extreme cases in 1965 were those with incomes in the $15-20,000 range, of which 63.9 percent reported a profit and were taxed accordingly, and the group of 766 farms with income greater than $1,000,000 of which only 14 percent reported a profit (U. S. Senate, 1968). Similar inequities in taxation can occur with property taxes based on assessed values of farm land. A 1975 study in Illinois showed a systematic bias in farmland assessment. Owners of farmland priced at $200 per acre paid taxes based on assessments of 38 percent of sale price. This tax rate is over three times that of farmland with a sale price in excess of $1,000 per acre. The latter was assessed at only 12 percent of sale price (*Successful Farming*, 1976).

Large farms also connote greater potential for environmental pollution, partly because of cheaper accessibility to operating capital and volume discounts on input commodities. Perelman and Shea (1972) cited data showing lower interest rates on loans for larger farms, and sizeable volume discounts on fertilizer (ten percent), insecticides (14 percent), and aerial crop dusting (25 percent) for farms of 3200 acres or more.

Energy

The fundamental task of any agricultural system—to harness solar energy for food production—is accomplished by using human, animal, and inanimate energy to channel the flux of solar energy in desirable ways. In chemistry a catalyst is a chemical that facilitates a reaction. Human and other energy forms have a similar catalytic role in agriculture. Because energy can be measured in universal units (calories, joules, BTU's, and the like), it has become fashionable to measure the productivity of an agricultural system in terms of energy output (of food) compared to total energy inputs, an approach that is somewhat misleading. Calories of crude oil energy are not calories of food energy, since the former have only a potential catalytic role, whereas the latter constitute energy consumable by man. Why, then, is energy analysis important?

First, energy is pervasive in the food supply system. The system uses energy to run farm machinery, to manufacture fertilizer, to ship both agricultural inputs and outputs, to produce agricultural chemicals, to produce machinery and building components—the list is endless. In the United States, 12 percent of our overall energy use is for agriculture (Hirst, 1974). Here, then, the food problem overlaps the "energy crisis," and both commodity and economic returns for energy inputs are of increasing importance. In the industrialized nations, marginal returns for increased investment in energy-dependent inputs are assessed with increasing care. Developing areas, where unit increases in energy-based inputs could have greater marginal returns in terms of food supply and human nutrition, must compete with the developed world in the marketplace for energy inputs.

A second factor is the interchangeability of land, labor, capital, and energy. Production otherwise increased by expanding land planted to crops may also be achieved by adding more labor, capital (such as irrigation systems), or energy (particularly as fertilizer). In the United States, capital and fossil fuels replace human labor and land, allowing the American farmer to produce fifty times more maize than would be possible by hand labor, investing only 22 man-hours per hectare of production compared, for example, to 1144 man-hours in parts of Mexico (Pimentel *et al.*, 1975:755). If energy inputs could be used to provide dry-season irrigation and fertilizer to produce an additional crop per year in rural Asia, Chancellor and Goss (1976) suggest that input of 0.453×10^6 calories of energy could produce the food needs of one person (0.952×10^6 calories per year in Asia) without any expansion of land. Thus, approximately three people could be fed with the catalytic energy role of one barrel of crude oil.

A final factor is that fossil fuel energy, as a major agricultural input, is a nonrenewable resource. To the extent that food production depends on energy subsidy, its energy needs must compete in the economic marketplace with other energy demands, and as part of the larger energy economy, it must eventually face the prospect of resource depletion. We must question the eventual fate of food supply systems based on a fossil fuel subsidy.

Estimates of the energy subsidy to the industrialized food system have been made based on food production alone and on the whole food supply system.[20] An example of the calculations possible for assessing the energy input and return from agriculture are average values for corn (maize) production in the United States (Table 20). Considerable technical detective work must be done to complete these estimates, since values must be ascertained for the energy costs of manufacturing machinery, producing pesticides, and running tractors, among others. The use of energy in corn production is worth pursuing because corn represents a middle level of production intensity, intermediate, for example, between vegetables and hay. Since 1945 the amount of energy contributed by labor has declined considerably, reflecting decline in farm labor populations over the same period. As the corn yield has increased over the same period so has the demand for all other inputs which utilize energy. Note the particularly large increases of fertilizer. This is to counteract the monocultural system's negative effect on soil quality:

> In 1939 corn was grown in three year rotation cycles of corn-oats-clover in order to regenerate the soil. No fertilizer was used, and 10,000 corn seeds were planted in an acre. The yield in the corn belt states was 38 bushels per acre. By 1970, however, rotation had been left by the wayside. Instead the average farmer employed 150 lb. of nitrogen fertilizer. Over 25,000 seeds were planted per acre and yields were 90-100 bushels per acre (Lerzu, 1974:8).

It is also apparent from the table that the energetic efficiency (the ratio between the amount of energy expended in production and the amount received from the harvest) of corn production is very low and has declined in the period considered. Thus, it was 3.70 in 1945 and only 2.82 in 1970.

[20] Works by Cervinka and colleagues (1974), Leach (1975), Pimentel *et al.* (1973, 1974), Steinhart and Steinhart (1974), Heichel (1973, 1974a, 1974b), and USFEA (1976) are worth pursuing in this regard.

TABLE 20. ENERGY INPUTS PER ACRE OF CORN PRODUCTION, 1945 and 1970

Input	Units	1945 Amount	1945 Energy Equivalent (10^3 calories)	1970 Amount	1970 Energy Equivalent (10^3 calories)	Ratio 1970/1945
Labor	hours	23	12.5	9	4.9	0.39
Machinery	Kcal × 10^3	180	180.0	420	420.0	2.33
Gasoline	gallons	15	543.4	22	797.0	1.47
Nitrogen	pounds	7	58.8	112	940.8	16.00
Phosphorus	pounds	7	10.6	31	47.1	4.44
Potassium	pounds	5	5.2	60	68.0	13.08
Seeds	bushels	0.17	34.0	0.33	63.0	1.85
Irrigation	Kcal × 10^3	19	19.0	34	34.0	1.79
Insecticides	pounds	0	0	1	11.0	∞
Herbicides	pounds	0	0	1	11.0	∞
Drying	Kcal × 10^3	10	10.0	120	120.0	12.00
Electricity	Kcal × 10^3	32	32.0	310	310.0	9.69
Transportation	Kcal × 10^3	20	20.0	70	70.0	3.50
Total Input		—	925.5	—	2,896.8	3.13
Corn Yield	bushels	34	3,427.2	81	8,164.8	2.38
Output/input ratio	—	—	3.70	—	2.82	—

The Kcal (kilocalorie) is the same as the food calorie.

Source: Pimentel *et al.* (1973), in *Science*, Vol. 182, November 2, 1973, pp. 444-445. Copyright 1973 by the American Association for the Advancement of Science. Reprinted by permission of David Pimentel and *Science*.

The example of energy input to corn production is paralleled by studies of other crops (Table 21) and by the total energy input for food production in the United States (Table 22). More elaborate machinery, fertilizer, and direct energy use are all important in the growth of energy subsidy for food production in the U.S. The food provision system, on the other hand, has used several times more energy than food production. For example, in 1940, over four times as much energy was used in food provision as production, falling to less than three times as much in 1950 and 1960, but rising to over three times as much in 1970. The system beyond the farm uses far more energy, and adds nothing to the amount of food that is available. Overall energy input in relation to agricultural productivity is illustrated in Figure 9, showing an asymptotic pattern. Note the increasingly marginal returns to increased energy inputs—as energy inputs are increased there are no longer correspondingly great increases in food outputs. To the extent that recent historical trends reflect productivity as a function of energy, small energy cutbacks would have little effect on productivity, but large cutbacks would apparently have a significant negative effect. Hirst (1974:138) has calculated the approximate energy use for food production among food groups in the U.S., suggesting that 43 percent of the energy is used for livestock products, 13 percent for directly consumed starchy staples, and 44 percent for other foods (including ten percent for alcoholic beverages). Reflecting on our earlier discussion of dietary composition this point must be underscored—we not only eat high off the hog, but in doing so we apparently use a significant proportion of our energy subsidy for agriculture. Figure 10 summarizes the picture for the industrialized agriculture of the United States, showing the amount of energy subsidy required to produce a unit of food energy.[21] For each food calorie a two-calorie energy subsidy produced the food,[22] and another six or seven brought the food from the farm to us.

TABLE 21. ENERGY EFFICIENCIES IN CALIFORNIA FOOD PRODUCTION[a]

Crop	Energy Output/ Energy Input Ratio
Barley	6.609
Wheat	5.363
Rice	2.554
Sorghum	2.534
Corn	2.311
Potatoes	2.119
Apples	1.267
Dry beans	1.150
Carrots	1.059
Grapes	1.054
Tomatoes	0.761
Peaches	0.731
Lettuce	0.337
Beans (frozen)	0.324
Cauliflower	0.248
Pears (canned)	0.241
Tomatoes (canned)	0.167
Broccoli (frozen)	0.133
Cauliflower (frozen)	0.123

[a] Farm production without processing, except as specified.
Source: Cervinka *et al.* (1974:113).

[21] This discussion for the U.S. has also been paralleled by studies at the state level and for other nations. For example, Cervinka *et al.*, (1974) have calculated that five percent of the California energy supply is used for agricultural production; Leach (1975:36) attributes 4.6 percent of total United Kingdom energy use to production, and 15.7 percent of that nation's total energy budget to the food system. Comparable figures for the U.S. are an estimated 12 percent of energy for the food system (Hirst, 1974) and an implied four percent for food production alone, using the variables in Table 22.

[22] Heichel and Frink (1975) calculated an energy subsidy of 4944 calories per day to produce a diet of 3391 calories, a ratio of 1.5 rather than the two implied by the Steinhart and Steinhart data (1974).

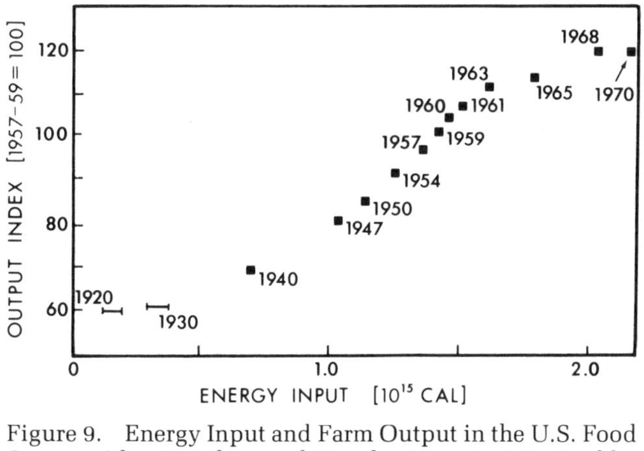

Figure 9. Energy Input and Farm Output in the U.S. Food System. After Steinhart and Steinhart (1974:310). Used by permission of John S. Steinhart.

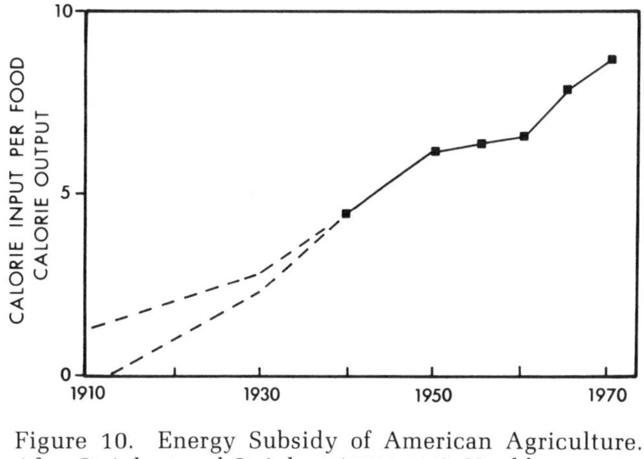

Figure 10. Energy Subsidy of American Agriculture. After Steinhart and Steinhart (1974:311). Used by permission of John S. Steinhart.

TABLE 22. ENERGY USE IN THE U.S. FOOD SYSTEM[a]

Component	1940	1950	1960	1970
Production:				
Fuel (direct use)	70.0	158.0	188.0	232.0
Electricity	0.7	32.9	46.1	63.8
Fertilizer	12.4	24.0	41.0	94.0
Agricultural steel	1.6	2.7	1.7	2.0
Farm machinery	9.0	30.0	52.0	80.0
Tractors	12.8	30.8	11.8	19.3
Irrigation	18.0	25.0	33.3	35.0
Subtotal	124.5	303.4	373.9	526.1
Processing:				
Food processing industry	147.0	192.0	224.0	308.0
Food processing machinery	0.7	5.0	5.0	6.0
Paper packaging	8.5	17.0	28.0	38.0
Glass containers	14.0	26.0	31.0	47.0
Steel cans and aluminum	38.0	62.0	86.0	122.0
Transport (fuel)	49.6	102.0	153.3	246.9
Trucks and trailers (manufacture)	28.0	49.5	44.2	74.0
Subtotal	285.8	453.5	571.5	841.9
Preparation:				
Commercial refrigeration and cooking	121.0	150.0	186.2	263.0
Refrigeration machinery (home and commercial)	10.0	25.0	32.0	61.0
Home refrigeration and cooking	144.2	202.3	276.6	480.0
Subtotal	275.2	377.3	494.8	804.0
Grand total	685.5	1134.2	1440.2	2172.0

[a] Values shown are 10^{12} calories.
Source: Steinhart and Steinhart (1974:309). Used by permission of John S. Steinhart.

Traditional Food Supply Systems

We now have an image of the American food system, productive and beneficent in some respects, wasteful and gluttonous in others. Borrowing a very effective technique used by Heilbroner (1963), let us try to envisage the food system of a traditional society on the fringes of development, in contrast with the U.S. pattern. Our example is chosen from among those hundreds of millions where traditional agriculture still persists. Let us begin with the farm itself, taking away 380 of the 390 acres, since 38 other families occupy that land. The buildings go as well, the barn, the implement garages, the corn crib, the silos, the milk house, the tool shop. In their place we have the eaves of the house and several mud-lined grain stores, a small corral for the oxen, and a perch for the chickens. Most livestock are gone, since there is not sufficient room for them on ten acres. Two oxen and a few chickens remain.

Farm machinery? It too has vanished, replaced by a small ox-drawn plow, a few odd knives, a machete, and a fine pair of hoes, one with a hand-forged blade, the other with a blade purchased in a country store. Obviously the gasoline and oil tanks are gone, with no machines to run. The file of old tax returns and farm records is now a tattered notebook with several important dates and tax receipts for a few years past. The diploma on the wall is now a third grade school certificate, and there are no files of old farm magazines.

The farmhouse? Remove the electricity, plumbing, floors, and all but two rooms. Replace two-by-fours and brick with poles and mud, windows with wooden shutters, the range with a small kerosene stove. There are no cupboards to be bare, and closets are pegs with a change of work clothes and one good garment for each family member. Hand woven baskets hold produce in the rafters of the house, and clay pots, filled at the stream a quarter-mile away, hold a day's drinking water supply. The water will be clear when the mud settles. Grass thatch or corrugated metal provides a roof, and during the rainy season drinking water will be collected from it.

You can guess what has become of the power mower, television, mixer, clothes washer, and other "necessities" of modern life. A battered bicycle, a transistor radio—these are the real luxuries. No driveways to shovel in winter, no air conditioning to break down in summer. But of course only a primary school nearby, a dispensary five miles away, a hospital 15 miles beyond, all over dirt roads without public transport. No grocery stores, shopping centers, or food stamps.

Modest production of cash crops means an annual cash flow of one hundred dollars, in addition to the cash value of the food crops (as if that is an adequate measure of their value!). Of the hundred dollars, ten may go to school fees, ten to taxes, twenty to fertilizer for the cash crop, twenty to laborers at critical periods during the crop year, twenty to household necessities like cooking oil, matches, needle and thread, ten to clothing.

Perhaps we belabor the image? Nevertheless, we must ask ourselves how a traditional system of agriculture functions, using the same frameworks we discussed previously. By looking at some of the impacts of the first salients of "development," we will also understand the enticing character of a transitional, "green revolution" farm, basic to understanding a third example of a food supply system which will follow.

Food Production

Traditional or preindustrial food supply systems may be defined as systems which produce food without the benefit of fossil fuel subsidy (Duckham and Masefield, 1970; Ruthenberg, 1971; Grigg, 1974). Our heuristic example is drawn from an area most familiar to the authors—tropical Africa—but this region differs substantially from Latin America, for example, in having fewer traditional cases of landlord-tenant relations, and from Asia in not having had widespread irrigation and its socio-political concomitants. Nevertheless, the conditions we describe are pan-tropical in distribution.

Food production in the traditional systems is built from a store of information—a genetic stock of crops and livestock, and human information that may be referred to as folk science or ethnoscience. Although individual differences in ability to know and teach about the environment occur in traditional society, there is little specialization with regard to agricultural knowledge. Virtually everyone has a detailed and intimate familiarity with environment, of an exceptionally high degree of ecological rationality as seen from our perspective (Conklin, 1967; Morgan and Moss, 1970; Knight, 1974; Berlin *et al.*, 1974). Space is allocated by traditional tenure systems, many of which are characterized by rights of use rather than by true ownership; land is allocated by the community for use, and is redistributed by the community as its membership changes. Land and crop rotation, burning, and, in some areas, recycling of animal wastes provide nutrients. In production systems with land rotation, known as shifting cultivation, planting is followed by long, wild fallow before clearing, burning, and recultivation. Natural ecological succession restores environmental productivity. More intensive land use, dependent on crop rotations and manuring, are characteristic of areas with higher densities of population (Boserup, 1965).

Solar energy in the traditional system is channeled, as in the industrial, through crops, but here the crops are often interplanted in the same field. Interplanting minimizes weed infestation, provides a less uniform environment for pests and diseases, and makes greater use of solar energy and soil nutrients. In many tropical areas, the "field" is a microcosm very similar in ecological structure to natural biological communities. Succession is controlled by the field preparation process (by machete and hoe, perhaps by animal-drawn plow), by weeding and cultivation as the crops mature, and by mixed cropping. Mixed crops, traps and snares for animal pests, flooding, guarding, companion planting (plants which resist pests are planted among the crops—as are marigolds and onions in the U.S.), and other techniques provide protection against pests and diseases. Harvesting is by hand labor or animal-powered machinery.

There are a number of spatial and temporal means for coping with seasonality and variability in a traditional production system. For seasonality, there are traditional calendars that control the scheduling of activities. The genetically-controlled phenology of crops also helps to match food production to climatic seasonality. Spatial dispersion of activities may extend the seasonality of crop production by taking advantage of different environments. If temperature and rainfall patterns are favorable, multicropping (using the same field for several crops in succession) provides a temporal sequence of food supply from the same field; grazing livestock on uncultivable, or fallow land also extends the period of food production. For pastoralists, seasonality may mean transhumance, seasonal migration of livestock that allows harvesting of solar energy by grazing in a number of places and environments.

The major response to variability in the traditional system is use of what Allan (1965) has termed the "normal surplus." Faced with environmental variability, traditional farmers in areas with adequate land resources typically cultivate more land or herd more livestock than would be required to produce adequate food yields in an average year. By cultivating more land—or cultivating land more intensively than usually needed—the farmer assures that although yields may be extremely low, enough food will be harvested even in the poorest years to supply the family until the next harvest. Clearly, this does not always work, for a new and unexpected threat to production might elude the normal surplus mechanism. For example, the normal surplus might be adequate for occasional drought, but not a locust invasion. The normal surplus strategy is so named because during all but the poorest years the farm family produces more food, often much more, than can be consumed. During the best years, some food is left to rot in the fields, feasting is common, and suitable foodstuffs are converted to alcoholic beverages. View the latter two, if you like, as a positive reward for the farmer working more than otherwise might have been required. In Africa, the normal surplus may have been the first commitment to a cash economy, since excess crops could be marketed once the economic infrastructure appeared.

In addition to the normal surplus mechanism,

farmers can also cope with variability by spatial dispersion of activities, including scattered fields and dispersed livestock. An example of the latter is the *tilia* exchange of cattle among the Pakot of Kenya, in which a person's cattle are dispersed among the herds of friends ostensibly as a binding of friendship, but also as a means to prevent disease, drought, rustling, or other hazards from decimating one's herd (Porter, 1965). Several of the seasonality- and variability-adaptive aspects of traditional food production also carry over into food provision.

Food Provision

Food provision in a traditional society is simplified by the maintenance of singular production-consumption units. Each producing unit (the family) lives on or near the farm, producing most of its own food needs and consuming much of what it produces. What transportation is necessary is provided by human porterage or animal energy. Traditional markets may take some food out of the local community or circulate food within it, but many of the food transfers would be by barter (for specialized services of a carver, physician, or diviner), reciprocity (sharing of labor, food, and other commodities on a reciprocal gift-giving and help-lending basis), and redistribution through traditional authorities (with taxation, personal wealth, or common fields providing food to needy families, as well as for common ceremonial occasions) (Polanyi, 1957). Food conversion is simple, with small amounts fed to livestock, some food converted to alcoholic beverages, and some moving through the mechanisms described to other consuming units. Dietary custom and etiquette govern food preparation and ingestion, as in the industrialized world. Food prejudices are common here as well as among ourselves, and some may have a rational basis (Simoons, 1967). For example, adult lactase deficiency means many of the world's people cannot digest milk (hence the intriguing title, "One Man's Milk is Another Man's Whitewash"—Harris, 1972).

In addition to the normal surplus, storage, migration, toleration of hungry seasons, redistribution, and dietary change can all be seen to meet the imperatives of seasonality and variability. In the overall food supply system, the traditional society's mechanisms for coping with the basic requisites of food production, seasonality, and variability are essentially local in scale. Although the complex network of flows and feedbacks that characterize the industrialized system are not present, traditional mechanisms meet many of the fundamental needs for food supply efficiently and resiliently.

Assessment

The characteristics of traditional systems include varying intensities of land use, particularly as a function of population density, and returns to labor are inversely proportional to land-use intensity. Boserup (1965) has suggested that at low population densities, traditional agricultural systems can achieve high outputs in proportion to labor input by using land rotation as a major structural element. As population increases, land must be used more intensively, and the required weeding and other labor to maintain productivity with increasingly shorter fallow means decreased productivity to labor, but increased productivity to land. The ultimate traditional end of this evolution is the irrigated, wet rice cultivation in southeast Asia, where marginal labor inputs may produce just enough food to sustain the laborer (Geertz, 1963). As a result of this kind of difference in land use and labor intensity, no single characterization of returns to land and labor is possible, perhaps with the exception that labor will be economized wherever possible—no one wants to work more or harder than necessary.

Traditional capital is obviously simpler than industrialized agriculture, often including hand-made implements. Complex irrigation systems are also capital, as are terraces and other land improvements. However, the terms of reference we have used before are useless in the traditional system. Capital investments must be efficient energetically, since they represent human and animal energy investment that must be produced by the system. Energy then is a relevant measure.

Studies of traditional food supply systems have been undertaken from an energetic viewpoint (Table 23). Values range from 5:1 (output: input) among Ugandan pastoralists to 65:1 among shifting cultivators in Africa. As would be expected, energy returns to energy invested must be significantly great to maintain the human community. It is equally important to recognize that these returns are accomplished without fossil fuel subsidy (Rappaport, 1971).

Technology is indigenous in the traditional food supply system. This need not mean that genetic and cultural information were developed independently within the local society, but that whatever the origin of these elements and whatever their technological products, they have been tested against local social and ecological environments, and modified and adapted through time (Janzen, 1973). That the system (and the society using it) persists is one indication of its success; another is our increasing appreciation of the complex cultural-ecological dynamics of such systems, and their resiliency to environmental stress.

We who are accustomed to industrialized agriculture often view traditional practices with wonder, if not disdain. Particularly vexing is the apparent inability of India to feed itself, while its over 200 million cattle are not consumed by the Hindu population. India's sacred cattle are an excellent topic to help us see unfamiliar practices in a more favorable light. Harris (1966) provided considerable insight into the taboo against beef consumption when he argued that the sacred cattle could be understood as

TABLE 23. ENERGETIC EFFICIENCY OF TRADITIONAL FOOD PRODUCTION

Society	Place	Energy Output/ Energy Input Ratio
Kalahari Bushmen	Southern Africa	7.8
Dodo Pastoralists	Uganda	5.0
Shifting Cultivators	Congo Basin	65.0
Tsembaga (shifting cultivators)	New Guinea	20.3
India (traditional)	South Asia	14.8
Rice cultivators	Southeast Asia	14.2–16.5
Chinese peasants 1935-37	Asia	41.1
Corn cultivation (hoe)	Mexico	30.6
Corn cultivation (hoe)	Guatemala	13.6
Cassava cultivation (hoe)	Tanzania	26.9
Cassava cultivation (hoe)	Congo Basin	37.5

Sources: Leach (1975); Pimentel *et al.* (1974).

ecologically rational. He suggested a symbiotic role of cattle with man, in which cattle provide milk, traction, and dung, plus beef and hides for Hindu untouchables, Moslems, and Christians. For Hindus, cattle are critical for supplying labor at seasonal bottlenecks in the staple grain economy. By using crop wastes and uncultivated land, they do not compete with man, and indirectly provide solar energy as a fuel (dung—two-thirds of which is used as the main source of domestic fuel in India) and organic fertilizer for fields. These observations do not prove that cattle numbers and products could not be used more efficiently, but they do suggest that behind seemingly irrational religious beliefs often lies an ecological rationality.

Harris's argument has been developed further from an energetic viewpoint by Leon (1975). In India, 29 percent of the matter provided to cattle is used, 22 percent of the energy, and three percent of the protein, in contrast to nine, seven and five percent in the United States, respectively (Table 24). Although the small proportion of human food provided to cattle in India could be directly consumed, Indian cattle provide food in excess of the edible food consumed, in contrast to the U.S. where six times as much edible food is fed to cattle as is obtained from them.

Finally, the traditional system's ecological impact has been evolutionary, rather than disruptive in nature. The theory of biological evolution concerns more than genetic material; it includes the evolution of interactive systems of living matter and environment. Thus the functioning and persistence of traditional systems may be seen as an enduring evolutionary product of human activity and environmental modification. We certainly cannot suggest that over a long period, without the Industrial Revolution, even nonfossil fuel subsidized agriculture may not have earned the accolade of "world food problem." However, the industrial revolution, growth of technology, and, for present developing areas, the colonial experience have brought a kind of "instant stress" to human experience. From the perspective of our recent history, we may see the rationality and ecological sensibility of traditional food supply systems. One of the dilemmas of development is the maintenance of the "positive" elements of tradition, while bringing change elsewhere: a process of "creative destruction" (Malassis, 1976). Planners usually neglect the former and the latter dominates, not always creatively.

Although our description of traditional food supply systems has been both brief and oversimplified, our intention was to invoke a feeling of appreciation for the positive aspects of these systems:

> . . . so called primitive societies developed technologies, techniques, and a store of practical knowledge of a wide range of sophistication, by what must be admitted to be the scientific method, and neither their accomplishments and skills nor those of societies *en voie de développement* should be ignored or discounted (Brown and Pariser, 1975:592-593).

When brought into contact with the industrialized world, traditional systems endure various kinds of stress which should be mentioned (Szentes, 1971; Porter and de Souza, 1974):

(1) Entry into the market economy, occasioned by taxation, conscription of labor, or the lure of commodities;
(2) Allocation of productive resources to that economy, through conquest by the colonial powers and European settlement, or allocation of indigenous agricultural resources (land and labor) to cash crops;
(3) Accelerated growth of population as a result of diminished effectiveness of biological and social controls, including declining rates of infant mortality, changing incidence of disease, and decreased warfare and social unrest;
(4) Imposition of various land management practices under colonial control, including

TABLE 24. INPUTS AND USEFUL OUTPUTS FROM U.S. CATTLE AND INDIAN CATTLE AND BUFFALO, 1972

Inputs and Outputs	Matter (10^{10} kg)		Energy (10^{12} calories)		Protein (10^{9} kg)	
	U.S.	India	U.S.	India	U.S.	India
Inputs						
Edible by man	11.9	0.68	38.8	1.7	16.0	2.1
Inedible by man	22.2	40.00	88.0	120.5	25.1	33.3
Total	34.1	40.68	126.8	122.2	41.1	35.4
Outputs						
Work	—	—	—	6.50	—	—
Milk	1.12	0.51	5.04	2.09	2.06	0.88
Meat	0.90	0.50	4.40	2.23	0.17	0.11
Hides	0.11	0.07	—	—	—	—
Manure	0.87	10.81	—	16.16	—	—
Total	3.00	11.89	9.44	26.98	2.23	0.99
Efficiency (%)	9	29	7	22	5	3

Source: Reprinted from "Agriculture: A Sacred Cow," by Bruce Leon, *Environment*, Vol. 17, No. 9, p. 38. Copyright © 1975, Scientists' Institute for Public Information.

cash cropping, prohibitions on traditional practices believed to be wasteful (such as shifting cultivation), and establishment of parks and reserves;

(5) Introduction of alien technology and ideas with both practical implications (new agricultural implements) and social implications (rejection of traditional knowledge and culture); and

(6) Decline of ecological conditions due to technological evolution insufficiently rapid to meet pressure on resources; to closing of such traditional outlets for man: land stress as territorial expansion; and to "careless technology" which has ignored local environmental milieus (Farvar and Milton, 1972).

In spite of this wide range of assaults, many traditional aspects of food supply systems persist within the "developing" societies. Not only do these systems continue to provide needed food supplies, but they have been able to withstand veneering by a developing cash economy. Indeed, some observers feel that small-scale, traditional farmers, responding to opportunities that are socially, economically, and ecologically rational from their perspective, are a positive element in rural development potentials in the nonindustrial world. Others would subscribe to the view quoted at the beginning of this chapter.

Agricultural Development—The Green Revolution

The "green revolution" began in the industrialized countries in the 1920's and 1930's with the breeding of new, high productivity crop varieties. Knowledge gained in the middle latitude areas was the basis for research programs in tropical areas, beginning with cooperative research by the Mexican Government and Rockefeller Foundation in 1943, and continuing today through an international network of agricultural research centers (Table 25). By the middle 1970's, a substantial proportion of wheat and rice production in developing areas was based upon high yielding or modern varieties (MV) of these crops (Dalrymple, 1974, 1975; Atkinson and Kunkel, 1976). The MV's helped to overcome a number of obstacles to improved crop yields, including environmental, cultural, and dietary problems with dissemination of crop varieties developed elsewhere; the limited response of traditional varieties to fertilization; the photoperiodic sensitivity to day length and the long maturing time which hindered multicropping; and potential lodging (flattening to the ground) of traditional tall varieties because of increased head size when fertilized (Brown, 1974:133-134). In the period from 1950 to 1970, grain yields of developing countries increased by almost one-third (USDA, 1974a:65). By 1973, approximately forty million acres were planted to MV's of wheat and rice in these developing areas (Dalrymple, 1975:19).

Recently, the International Rice Research Institute (IRRI, 1975) coordinated a study of 36 villages in rice growing areas of Asia to examine the extent to which modern rice varieties had been adopted by farmers in areas where inputs and markets were reasonably accessible, but where intensive campaigns to introduce the modern varieties had not occurred. Also examined were economic returns to farmers, changes in income and expenditures, and use of other items of the MV package—fertilizer, planting methods, and chemicals for pest and disease control. Although it is erroneous to generalize over a widely dispersed sample of farm villages, it is useful to list some general observations from this study:

(1) A majority of farmers had tried MV's, and it is reasonable to assume all farmers are aware of them;

TABLE 25. THE NETWORK OF INTERNATIONAL AGRICULTURAL RESEARCH

Institution	Location	Founded	Research Interests and Regional Coverage
International Rice Research Institute (IRRI)	Los Banos, Philippines	1959	Rice—worldwide with emphasis on Asia
International Center for the Improvement of Maize and Wheat (CIMMYT)	El Batan, Mexico	1964	Wheat, maize, barley, triticale—worldwide
International Center for Tropical Agriculture (CIAT)	Palmira, Colombia	1968	Beans, cassava, maize, rice, beef—worldwide for lowland tropics, with Latin America emphasis
International Institute for Tropical Agriculture (IITA)	Ibadan, Nigeria	1969	Root and tuber crops, grain legumes, cereals—worldwide for lowland tropics with African emphasis
West Africa Rice Development Association (WARDA)	Monrovia, Liberia	1971	Rice—West Africa
International Potato Center (CIP)	Lima, Peru	1972	Potatoes—worldwide
International Crops Research Institute for the Semi-Arid Tropics (ICRISAT)	Hyderabad, India	1972	Sorghum, millet, legumes—worldwide, unirrigated semi-arid tropics
International Board for Plant Genetic Resources (IBPGR)	FAO, Rome, Italy	1973	Conservation of plant genetic material—worldwide
International Laboratory for Research on Animal Diseases (ILRAD)	Nairobi, Kenya	1973	Livestock diseases—Africa
International Livestock Centre for Africa (ILCA)	Addis Ababa, Ethopia	1974	Livestock production—Africa
International Center for Agricultural Research in Dry Areas (ICARDA)	Lebanon	(planned)	Wheat, barley, legumes, oilseeds, cotton —worldwide with emphasis on semi-arid zones with winter rainfall

Sources: Wade (1975a:587); Jennings (1976:188).

(2) The use of fertilizers, insecticides, herbicides, and tractors preceded the introduction of MV's in many areas, and fertilizer use, in particular, was practiced by a majority of farmers before MV technology. A Pakistani village is an example (Figure 11);

(3) Asian rice farmers are not resistant to change, although adopters usually did better than their neighbors even when they grew local varieties, suggesting that they were better farm managers;

(4) In many cases the MV's did not provide greater yields than local varieties, but a shorter growing season and nonphotoperiodism means two crops per year are produced, with some areas growing five or six crops in two years;

(5) Where MV's made a second rice crop possible, it frequently displaced vegetable and pulse (legume) crops, with a potential for decreasing the quality of family diet, depending on level of food expenditures;

(6) The MV's did not decrease labor input, but increased labor requirements even in areas with tractors;

(7) Nevertheless, agricultural laborers seldom experienced the same change in economic standard of living as experienced by farmers who adopted MV's, although only about two-thirds of MV farmers reported increased profits and one-third increased level of living (Table 26);

(8) Although increases in level of living and in profits were not confined to landowning farmers with large holdings, increases were concentrated among this group, who also had more ready access to farm inputs and credit, and who had typically adopted MV's earlier;

(9) Typical kinds of expenditures by MV farmers include food, clothing, education, housing, medical care, bicycles, radios, other appliances and agricultural investments such as tractors and wells;

(10) Many constraints to increased production of MV's are those associated with the package—pest and disease problems; fertilizer availability; adequate irrigation and drainage; availability of credit; availability of seed;

(11) Local varieties are still important, both for disease resistance and for local foodstuffs; MV's are usually the marketed variety if both are grown, but, in at least one government,

price support of traditional export varieties meant lower MV acceptance.

This summary, as well as the detailed study, suggests very strongly an evolutionary rather than a revolutionary change in association with modern varieties.

In Java, two villages studied showed a marked contrast in acceptance of MV's of rice. In the East Java village, MV acceptance was high (99 percent), as was population density (2137 people in 453 households on 170 hectares of farmland). The West Java village had a population of 3322 people (958 households) on 480 hectares of farmland, but low acceptance of MV's. Both areas had access to improved local varieties before MV availability. MV's were introduced to each village in 1968, and first planting was undertaken by village leaders. In East Java, acceptance was rapid and remains high, whereas in West Java, pest infestation caused severe crop loss and return to local varieties. In the latter village, local improved varieties outyielded MV's, whereas in East Java the MV had marginally higher yields, in spite of a marked increase in fertilizer use. Acceptance appears to be more related to shorter maturity, critical in an area of intensive double cropping. In East Java, MV's required more labor than local varieties, primarily as a result of weeding required by the high fertilizer inputs which encourage weed growth. In West Java, harvesting the MV's took less time, because the sickle rather than the hand knife was used. In both areas, the rice crop is very labor intensive (147-276 man-days/ha) with double the labor input typical of Philippine farms (70-130), for example. Both areas experience annual rice shortages, and evidence suggests that even in the MV area of East Java, farmers are no better off after having accepted MV's (Table 27). The study concludes that, "there is no sign yet of any 'green revolution' in the sample areas" (IRRI, 1975:199; see also Franke, 1974).

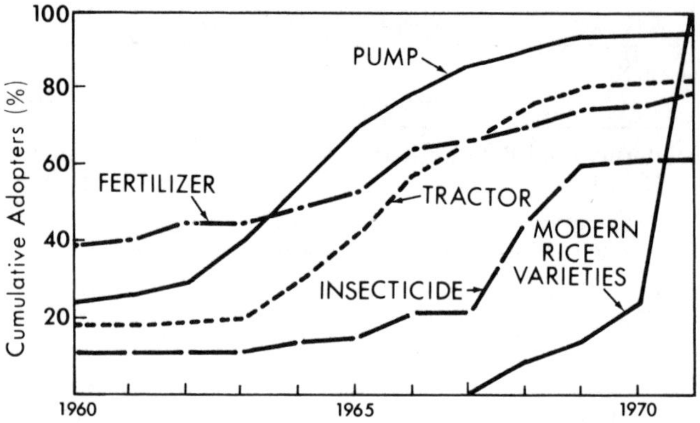

Figure 11. Adoption of Improved Farming Practices in Aroop, Pakistan. After IRRI (1975:234).

Given the present importance of the green revolution technological package for increasing food supplies, we shall look at dimensions of this technology from the point of view of vulnerability. "World food security" commonly refers to problems of greater concentration of surplus food production in limited areas and the absence of a cushion of food stocks or land reserves to meet needs created by extraordinary events (USDA, 1974a:40-47). However, the diffusion of technology from an industrial base poses problems that may be more localized—at the scale of a nation, region, or even farm. The concept

TABLE 26. INCREASES IN PROFITS AND LEVEL OF LIVING IN SELECTED AREAS WITH MODERN RICE VARIETIES IN ASIA

	Percentage of Farmers Reporting					
	By Tenure			By Farm Size[b]		
Growers of	Owners	Tenants	All Farms	Less than 4 ha	4 ha and over	All farms
			Increase in Rice Profits			
Modern varieties only	59	33	50	50	87	82
Some modern varieties	82	33	78	50	75	59
combined[a]	73	33	65	50	84	73
Local varieties only	20	1	12	6	6	6
all farms	44	10	32	26	66	44
			Increase in Standard of Living			
Modern varieties only	27	42	32	25	47	44
Some modern varieties	44	33	43	21	38	27
combined[a]	38	40	38	22	45	38
Local varieties only	6	6	6	2	17	6
all farms	21	14	18	8	36	21

[a] Value for all farmers who grew any modern varieties.
[b] Owners only.
Source: IRRI (1975:354).

TABLE 27. LENGTH OF FAMILY FOOD SHORTAGES IN AN EAST JAVA VILLAGE BEFORE AND AFTER ACCEPTANCE OF MODERN RICE VARIETIES

Year	No. of farmers	Percent Use of MV	Duration of Food Shortage (months) (Percent of farmers reporting) 0	1	2	3	4
1968/1969	70	40	35	3	20	38	4
1971/1972	70	92/94[a]	33	5	20	39	3

[a] Dry season/wet season.
Source: IRRI (1975:196).

of vulnerability was developed by Sprout and Sprout (1974). Following their typology, food supply vulnerability may arise from events in either the physical or social environment, generated by disruption within the local area or by actions originating from abroad. We will explore some aspects of new vulnerabilities introduced by the green revolution in reference to some basic requisites for a food supply system.

The use of imported green revolution techniques creates radical changes in the distribution of technical and genetic information. Technical information is specialized knowledge whose application allows use of sophisticated production procedures. Importing technology involves far-reaching changes in both the distribution of knowledge and the process by which information is disseminated throughout a society. Such changes are imposed upon societies with varied characteristics. Some impacted areas, for example the original sites of commercial exploitation in Mexico, remained traditional until their take-over by the Mexican government (de Alcantara, 1973-74). Elsewhere, the IRRI study (1975) noted that prior to the green revolution in Pakistan, imported commercial infrastructure and many green revolution innovations were already well-established. Information flow can be seen in international scholarships, the focal role of the agricultural research network, and training institutes run by multinational agribusiness corporations. A spirit of elitism develops among the modern agents of change, with risk of losing traditional knowledge. Rao (1974) has noted that extension agents fail to include farm input in information flow, so that extension is a one-way flow, emphasizing farmers who are identifiably progressive. In addition, other costs are sustained in acquiring green revolution information, including foreign exchange losses (patent and licensing fees, payments for expatriate expertise), the brain drain to developed countries, and potential inapplicability of particular technologies for recipient environments. We judge the loss of local environmental knowledge, as well as of local genetic material, a vulnerability, since these sources of future adaptations may no longer be available.

The seeds produced each year by food crops contain genetic information. In traditional systems either the individual farmer collects and stores this information or limited local trade occurs. This system thus assures that seeds adapted to each agroecosystem will be available. When imported seeds are distributed, other varieties formerly used are rejected in favor of the high yielding varieties. The loss of genetic variability is felt by some ecologists to increase the vulnerability of physical systems (Dasmann, 1973). Whether this is the case, such a change does alter accessibility to genetic information from a ubiquitous to a commercial system.

Counteracting the loss of genetic information is the development of regional, national, and private stores of germ plasm. However, such peaks in a nation's information surface are very vulnerable to unpredictable events, such as environmental hazards, and to social disruption, such as inadequate funding or even sabotage. Genetic erosion has been characteristic of Europe and North America for some time (Miller, 1973; Harlan, 1975). The urgency of the green revolution problem is that the new varieties are increasingly being exported to areas which were the original genetic sites for the major world food crops, and from which loss of genetic information would be particularly unfortunate (Frankel, 1973; Oldfield, 1976).

The introduction of high response varieties of seeds has an impact on the provision of land with several different implications for food system vulnerability. The importance of access to commercially-sold inputs for MV success gives established landowners with surplus income an advantage over smaller farmers. This initial advantage is translated partially into new land purchases and increased concentration of land ownership. Since ownership is already highly concentrated (Griffin, 1972; Rao, 1974), this increased social inequality can only increase the potential for vulnerability to social disruption. In addition, like large farms in the U.S., large farms in developing countries have financial incentives to use potentially harmful levels of agricultural chemicals (Cleaver, 1972).

Water is a critical element for success of MV's. The FAO projects that by the year 2000 controlled water use will increase by 240 percent in the agricultural sector (Brown, 1974:101). Institutional problems of water management, declining ground water tables, silting of reservoirs and canals, and energy demands for irrigation are all sources of food system vulnerability.

The green revolution brings an important change in the pattern of nutrient control used for agriculture. Manufactured sources are supplying increasing shares of nutrient supplements, a transition encouraged by the use of MV seeds. The new seeds require assured quantities of nutrients in concentrated amounts and this means use of commercial fertilizers which until recently were cheap. Such a change has several implications for food system vulnerability. First, vulnerability may occur because of price or supply changes for fertilizer. Yield potentials of MV's, highly dependent on fertilizer inputs, may be threatened. Second, inequities in fertilizer distribution at an intra- as well as international scale mean that fertilizers are not allocated where marginal productivity is greatest. The opposite appears true—the wealthy have access to fertilizers at levels giving little marginal returns. Third, fertilizer production is often located outside national boundaries, with resultant vulnerability to political or economic disruption.

The green revolution technology replaces traditional methods of weed, pest, and disease control. For example, in some rice producing areas, a combination of deep flooding and fish introduction were used as insect controls. MV plants are usually short-stemmed to prevent lodging, thus precluding deep flooding; persistent insecticides kill fish and diminish the availability of this protein source (Palmer, 1972). Monocultural practices, particularly region-wide planting of identical crop varieties, impose the threat of epidemic diseases, and the emergence of insecticide-tolerant pest species.

Agrichemical residues in the diet also affect the health of the population, especially that of agricultural workers. Lakshminarayana and Menon (1972) noted that in India the highest agrichemical concentrations are in starchy vegetables and cereals, thus having a greater impact on the poor who consume more of these foods. Rao (1974) noted the increased dangers of pesticides for users with low skill levels, such as those in developing areas. The danger of exposure to toxic chemicals is greatly affected by the accuracy with which instructions are followed, provided one can read.

Returning to the Sprouts' typology, we can see that the green revolution brings with its benefits a variety of vulnerabilities. Among physical vulnerabilities are those associated with local environmental disruptions of supply linkages for needed inputs and increasing international dependence on fossil fuel subsidies in the face of depletion of this resource. In addition to international social disruption affecting input supply lines, internally the green revolution may exacerbate status distinctions based on education and wealth, with the concomitant risk of increasing internal unrest.

Perhaps the most important aspect of the green revolution approach is the removal of the multiplicity of traditional means for coping with variability, without gaining the institutional, technological, and spatial processes that buffer the industrialized agricultural system against variability. A new set of dependencies is created for the farmer and consumer and should those dependencies fail, traditional means may no longer be able to operate.

The gains of the green revolution are nevertheless important. First, it has contributed at least to maintaining food supplies for growing populations in developing areas. Second, it has demonstrated the receptivity of traditional farmers to innovation, provided the necessary institutional elements are present (Crosson, 1975). Third, it has given developing countries a greater period in which to address population problems (Brown, 1974:145). However, the model of industrialized agriculture implicit in the green revolution is certainly a questionable solution to the world food problem.

Common and Contrasting Elements

All food supply systems must overcome spatial and temporal tensions between the photosynthetic process, harvest, and consumption. In doing so, all follow a basic set of requisites, some addressed simply and others dealt with in a high degree of complex elaboration. Both traditional and industrialized food supply systems are the result of evolutionary processes, and incorporate a variety of mechanisms for coping with seasonality and variability to assure food availability. Although the green revolution farm can be seen as evolutionary in the context of a longer period of intrusion of industrialized agricultural practices into nonindustrial areas, all these practices are alien in the context of developing world societies. Producing more food from the farm does not solve the plethora of other problems of development—including distribution of food equitably according to needs. The commitment to incipient industrialized agriculture imposes vulnerability to events beyond the control of the developing nation.

The green revolution can be seen primarily as an effort to bring the genetic and fossil fuel subsidized technology of the industrial world to developing areas in the hope of substantially increasing agricultural output per unit of land. Although it is desirable that the yield gap between the developed and developing areas be narrowed to bring food more equitably to the world's people (Table 17), it is uncertain whether the financial resources will be available, and whether earth resources will permit the developing areas to "catch up." In our view, narrowing the yield gap may require the willingness of industrial areas to tolerate declining yields.

The green revolution does not appear to have generated the numbers of displaced, landless people once expected, but, in spite of a continued high labor requirement, similar to if not greater than that of traditional agriculture, it has generated labor problems. One of these is seasonality of labor demand and a consequential lack of steady, reliable employment (Feder, 1973-74). Potential labor exploitation may result, as well as an impetus toward mechanization of labor bottlenecks. This then leads to the question of technologies appropriate in design and scale to developing areas, and thus to the role of agribusiness in development. A multinational corporation's primary concern is profit (Barnet and Muller, 1975), and profit will be maximized by sale of existing technologies whose research and development costs have already been recovered in home markets. Officials of developing nations, trained as industrialized agriculturalists, understandably follow FAO leadership in viewing mechanization and use of chemicals as indispensable for agricultural development. The FAO, in turn, draws upon its Industry Cooperative Program for advice. The ICP has 100 multinational agribusiness firms who have joined with FAO to accelerate development cooperatively (DeMarco and Sechler, 1975:75-77). The pressures toward an industrialized prototype for agricultural development are obvious.

For all food supply systems, spatial organization is a fundamental approach toward meeting the various requisites we have listed. Indeed, it is spatial organization that creates the fact of a world food system, and therefore a world food problem. Let us begin at one end of the scale. For the individual farm, microspatial organization of the field is an important element in harvesting solar energy resources, and in traditional agriculture it has been a major element in meeting succession and protection imperatives. Also, the organization of the farming unit affects its productivity. Traditional farmers manage their easily accessible dooryard gardens more intensively than distant fields. For industrialized farms, efficiency of operations and equipment performance is affected by farm layout; as in traditional agriculture, spatial dispersion of activities may function as a device for buffering the system against variability.

When we move up the scale from the farm, we leave many traditional systems behind, although some, like Asian paddy cultivation, depended upon elaborate social, political, and resource management. It is in the industrialized and green revolution food systems, however, where spatial organization is indispensable at regional, national, and international scales. Flows of agricultural inputs to the farm, and flows of commodities from the farm and eventually to the consumer constitute a major element in a society's basal metabolism. A multiplicity of linkages for any one kind of flow, and the capability of tapping a wide spatial network of resources, give industrialized agriculture protection against variability that is absent from green revolution situations. We should not be deceived by the stability that comes from this spatial redundancy in industrial agriculture—ultimately we depend upon fossil fuel subsidies, and when the well runs dry, the system will no longer be energized.

What is the world food system? It is the combination of material, energy, and information flows that disseminate the impacts of actions that occur in parts of the food supply systems we have discussed. Prices are perhaps a most obvious example: food prices in developing countries are linked to world energy, fertilizer, and food prices without regard to the productive capacity of the country itself (Chancellor and Goss, 1976:215). Suppose a country imports half its fertilizer supply, farmers provide twenty percent of food needs by growing green revolution crops, and half the nation's families live on farms. Consider a doubling of imported fertilizer prices. Limited foreign exchange means fertilizer imports must be halved, and green revolution farmers bid up the price of local fertilizer, limiting its availability to others. Marketable crop surpluses drop, particularly if farm families consume their normal needs and market only what is excess. Market shortages induce buying of food supplies in anticipation of price increases (hoarding), which further limits supplies and drives prices upward.[23] People with low income are forced to starvation level diets, requiring the government to import and sell food at artificially low internal prices. These purchases limit new energy resource and fertilizer plant development, because they exhaust foreign exchange reserves. Thus the opportunity to equilibrate the original price perturbation internally is prohibited. Such an example demonstrates that with agricultural development based on foreign technology, energy, and material inputs, world markets and prices effectively penetrate a country, an obvious vulnerability.

Since the industrialized nations effectively tap most of the world's natural resource markets and control the provision of industrialized agricultural inputs, the world food system is in fact created by that group of nations. Not only are the major surpluses of agricultural commodities controlled by developed nations, but so too is the food production capacity of developing nations. The latter must shop in industrial marketplaces for agricultural inputs and compete with the developed nations for energy resources. The substance of the world food system, like the industrialized food supply system, is becoming increasingly concentrated, both politically and in the emerging role of multinational corporations operating in developing areas:

> The role of the multinational corporation as world distributor is most dramatic in the area of food. Agribusiness is now buying or renting more and more arable land. Decisions on what to plant and where to distribute the harvest are made with the balance sheet in mind. Thus it is profitable in poor countries

[23] This is a good example of a self-fulfilling prophecy. People expect food price increases; some buy large quantities based on that assumption; further shortages occur, driving prices upward; thus the assumption of price increases is proved true, when such increases may have been created (or accelerated) by early buying and hoarding.

> to use land for exportable luxuries even while the people are suffering severe malnutrition because it does not grow enough grain (Barnet, 1976).

The developed nations have assumed some humanistic responsibility for addressing the world food problem:

> . . . we . . . proclaim a bold objective—that within a decade no child will go to bed hungry, that no family will fear for its next day's bread, and that no human being's future and capacities will be stunted by malnutrition (Secretary of State Henry Kissinger to the World Food Conference, 1974).

Such a commitment must be more than rhetoric. A pledge toward solution of the world food problem by the industrial nations is proper and correct: increasingly we *are* the world food problem.

III. SOLUTIONS

Perhaps it would be better to recognize and acknowledge forthrightly that the rest of mankind will probably never consume as much food as the average American.

L. R. Brown (1974:44)

What are the alternatives for solving the world food problem? For the optimist, science and technology, paralleled by application of demographic brakes in developing areas, promise an end to hunger and famine. Technological optimism sometimes denies even the necessity of abating population growth:

> At present [1967] there are about 3 billion men living on earth. The predictions are that there will be 6-7 billion around the year 2000. At this rate of increase, the number of 100 billion will be reached in 200 years. At that time starvation may be a bitter memory only, but today many, many persons go hungry in a world where the technical ways and means to prevent this are available (deWit, 1967:320).

For the pessimist, the future holds the promise of an increasing gap between hungry populations and the ability of the world to feed them. To some observers, the world's population has already exceeded the earth's carrying capacity (Whittaker and Likens, 1975). To others, we face the dilemma of providing food aid as a temporary palliative, only later to witness more massive starvation among even larger population numbers:

> There can be no moral obligation to do the impossible . . . At some point, we in the United States are going to find that we cannot provide for the world any more than we can police it . . . our position is this: The sovereign right of each nation to control its own reproduction creates the reciprocal responsibility to care for its own people . . . we must not permit our aid to underwrite the failure of some nations to take care of their own (advertisement by The Environmental Fund, 1976).

The alternatives then are those of triage—to help those nations where technological development is likely to be accompanied by population control, and to abandon the world's "basket cases," those who would swamp the lifeboat.

To help us understand the major directions that solutions to the world food problem might follow, it will be useful to categorize a number of contrasting viewpoints. Earlier we conceived of various definitions of the world food problem from a spatial perspective. This grouping was particularly useful for discussing food supply systems in industrialized, traditional, and transitional settings. We noted a wide range of problems and potential solutions in each setting. Now, we will take the same kinds of problems definitions and cast them in a different way, treating them topically rather than spatially. Figure 12 suggests problem clusters seen from a topical viewpoint, including views of the world food problem as fundamentally an economic, technological, environmental, demographic, or moral dilemma. Among economic problem definitions are questions of supply and demand imbalance; technology includes the need for research and development of new genetic materials and production technologies; environment includes problems of ecological degradation and environmental change; demography deals with the question of human numbers; and morality or equity includes the problem of unequal access to resources, dietary extremes, and the fundamental structure of the world economic system.

For any analysis, the problem is typically conceived from one or more viewpoints on its causes, followed by solutions to these causal factors. We view solutions in three broad groupings: those focusing on technology, population, and moral equity. In general, technological solutions are those which aim to enhance industrialized agriculture at home and propagate it in developing areas, along with the socioeconomic concomitants required to diffuse technological innovations. The demographic cluster focuses ultimately on population numbers, for even the most optimistic technologists nevertheless concede that the earth is a spaceship, with some upper limit on human numbers. Clearly no analyst would be so naive as to assume that abated population growth alone will solve the world food problem, but few solutions omit an implied, if not explicit, population component. The equity cluster is a bundle of solutions that are essentially political, requiring commitment to a fundamental reordering of food supply systems at all scales, accompanied in large measure by a similar reordering of world economic processes. Implicit in these solution clusters are alternative futures, ranging from an integrated world food system of high technological refinement to a reassertion of local adaption and initiative, accompanied by equitable access to world resources. Between these lie two forms of inequality and differentiation—duality and triage. Duality implies

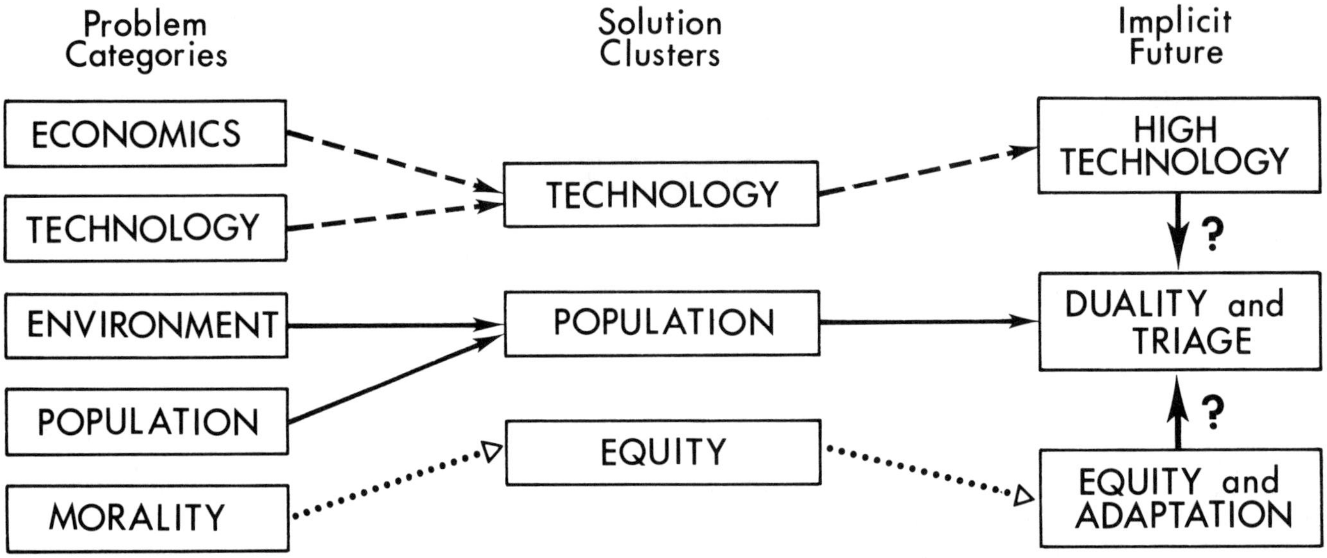

Figure 12. Problems, Solutions, and Implicit Futures in Views of the World Food Problem. Arrows suggest some among many of the possible analytical paths that may be followed.

marked local differences in food production technology, social status, and wealth; triage implies the same dimensions among rather than within nations, the perpetuation of the "haves" and "have-nots," of the superrich and the hopelessly poor.

In this section, we first want to suggest that the considerable disagreement as to the nature of the problem and its solutions reflects positively on opportunities for solving the problem! Then, we will discuss briefly the three solution clusters, finally concluding that whereas duality and triage are probable future courses for mankind, this need not occur if equitable and humane commitments are made by the citizens of industrialized nations.

The Adaptive Man-Environment System

In the context of this paper an adaptive system refers to the ability of human society to interact with its environment to produce food. An adaptive system has several fundamental characteristics. First, it must be linked to its environment, in this case by flows of matter and energy. It must be able to extract information from environment, and to create new structures and behaviors in response to environmental changes. These changes may come from within (innovation) or without (diffusion). Then, processes of selection must exist, either in society or in environment, to test potential new behaviors. Finally, a memory must exist to preserve successful adaptations for future use (Buckley, 1968).

The biological mechanism of evolution is one example of the adaptive systems framework, where new behaviors (mutations) apparently are randomly generated without regard to environmental adaptivity. Genetic structure remembers changes that have survived selection mechanisms, and reproduces them. In the process of cultural adaptation, while not denying the possibility of an "out-of-the-blue" act of creativity, the creative process is guided by information flow from environment. Knowledge and the enculturation of children are means by which cultural memory is preserved. Thus both genetic and cultural adaptation are possible in interaction with environment (Figure 13), and both are relevant to this discussion. Genetic adaptation is an important element in providing new crop and livestock varieties to meet imperatives of food supply, whereas cultural adaptation may be seen from a wider perspective as a process by which mankind solves its problems.

How does cultural adaptation occur? One useful perspective on that process is to think of certain filters through which new ideas pass. Cultural filters measure social acceptability as well as test new ideas with respect to existing knowledge. Economic filters test the viability of a new idea from an economic viewpoint, most obviously in monetary returns to investment but, in broad terms, as a measure of desired returns in comparison to those from alternative investments of scarce inputs. Ecological or environmental filters take the test to the real world, and in that context select for viability (Firey, 1960). These perspectives, however, do not tell us where ideas come from, nor how they relate to the contemporary situation.

Many ideas are derived in relationship to ongoing problems or to salient issues of the moment. Figure 14 schematically suggests a sequence of events that occurs in society's problem solving sequence (Ormrod, 1974). Awareness of the problem comes from a linkage between society and environment. The creation of new behaviors includes analysis and eventual definition of problems, and the search for solutions that are put forward and culturally screened. Then,

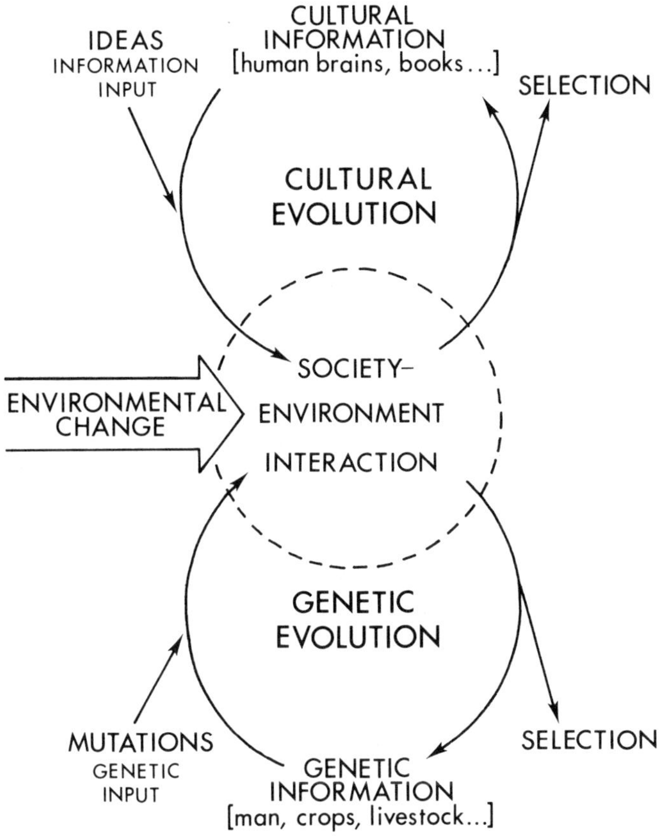

Figure 13. Information, Adaptation, and Continuity in a Man-Environment System. Modified from Bajema (1972:225). Used by permission of *Social Biology*.

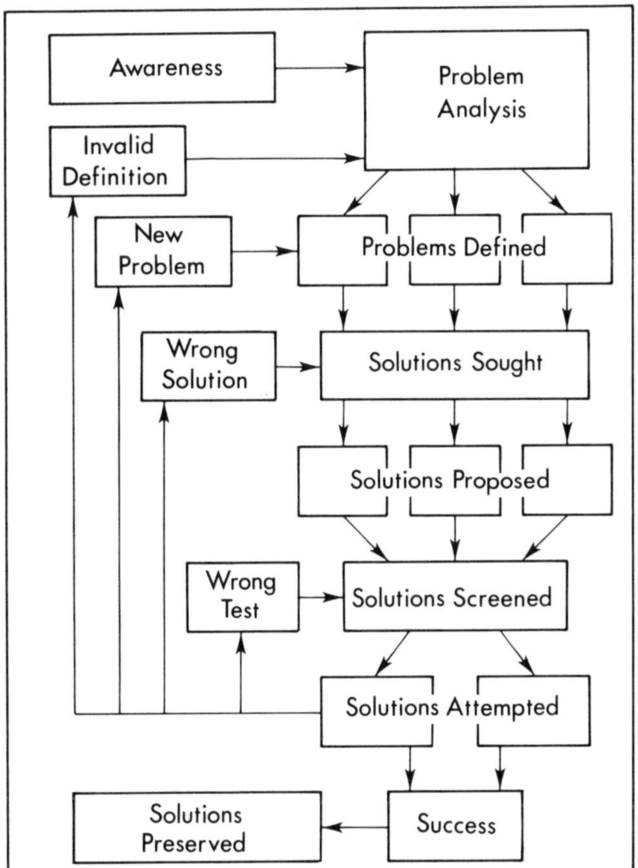

Figure 14. Problem Solving in the Adaptive Human System. After Ormrod (1974:231). Used by permission of Richard Ormrod.

attempts at solution may have several results, with successful solutions preserved (culturally encoded). Preserved knowledge in traditional societies is referred to as ethnoscience, whereas in modern societies knowledge is science and technology.

In the adaptive systems framework, the search for solutions is seldom random, but is frequently channeled by our understanding of cause and effect. We accept that events have causes and seek to understand the identity of causes, the mechanisms by which they operate, and the processes they represent. For example, searching for a solution to the problem of increasing crop yield depends on understanding what it is that makes plants grow. Plant growth may be limited by insufficient water, infertile soil, competition from other plants, diseases, or pests. If you are a farmer in southern Nigeria, you may understand the use of chemical fertilizers because your traditions have made you aware that plants grow by taking up substances ("fat") from the soil. Only if you are an agronomist are you likely to understand fully that some crops are producing their maximum yield as constrained by their genetic endowment. The Nigerian's search may be for commercially produced "fat" for the soil; the agronomist's for genetic material with a greater efficiency of solar energy conversion and fertilizer utilization.

Three important implications follow from an adaptive systems perspective on the world food problem. First, it is important to recognize the inherently conservative nature of adaptation. Perhaps in biology you heard the phrase, "ontogeny recapitulates phylogeny." This means that the embryonic development of advanced life forms follows an evolutionary-like process, progressing from single cell to increasingly complex forms as it grows in the egg or uterus. Genetic evolution builds upon itself, rather than totally restructuring life, and accumulates minor modifications which have permitted survival in different and changing environments. In the case of society's adaptation to environment, even adaptation may suggest too extensive a change. A better term might be adjustment, a series of minor changes that accumulate over time, but which when viewed from only two ends of the time spectrum might appear to be a more radical adaptation. Adaptation does occur, however, when dissonance between society's understanding of and behavior in environment utterly fails, and when survival itself is threatened. The temporary abandonment of pastoral life for urban refugee camps in the West African Sahel represents this kind of adaptation, at least in the short run.

The conservative nature of adaptation can be seen in the process by which the world food problem is addressed. Most solutions represent a cumulative tinkering with the status quo, rather than any fundamental reordering of the system as it now stands. Only massive dissonance could alter this pattern, but for some observers such dissonance appears at hand.

The second implication of the adaptive systems framework is the value of a multiplicity of problem definitions, solutions, experiments, and eventual answers. The greater the universe of perspectives, the greater the opportunity for productive solutions to be found. Rather than experience frustration at the complexity of the world food problem, we can be encouraged by the plethora of analyses and approaches being undertaken. This further implies that we must avoid two potential reactions—dependency on a monolithic response to the problem (particularly, putting all the eggs in the industrialized agriculture basket), and feeling a hopelessness that causes us to continue as in the past, becoming insensitive to the problem itself. The concentration of technology and economic power in the industrial world and in a small number of powerful firms within it enhance the risk of both reactions.

Finally, the adaptive systems perspective helps us to link problem definitions, suggested solutions, and implicit futures. We can trace one or several solutions proposed in relation to conceptualizations of the nature of the problem, as well as the local and global implications of the solution (Figure 12). For example, if one views the problem narrowly as excessive population growth in nonindustrialized areas which is hopelessly destroying the ability of resources to provide sufficient foodstuffs, triage is an implicit (if not an explicit) future should population control fail to materialize. The greatest impediments to solving the world food problem may well be ill-conceived definitions of its causes, leading to misdirected solutions.

Technology

Much of the contemporary literature on the world food problem focuses on technological approaches to increasing food production, as well as on the planning necessary to disseminate technological breakthroughs and encourage technology dependent agricultural development (*Science*, 1975; *Scientific American*, 1976). For this reason, we will consider alternative approaches in greater detail, briefly suggesting major directions of technological emphasis and focal questions concerning technology based on agricultural experience in the industrial areas.

A wide range of research directions focuses upon technology (Table 28) as a solution to the world food problem. Many approaches represent a continuation of emphases already evident in the green revolution, including crop and livestock genetics, farm operations, and increasing reliance on institutionalized research. A major emphasis is on direct and indirect energy subsidies to agriculture. Mechanization, agricultural chemicals, and irrigation represent the substitution of energy for land and labor; maintenance of food stockpiles and commodity food aid to abate famine are similarly energy dependent. It is obvious that it would be impossible to produce all the world's food by energy intensive industrialized agricultural systems. To increase energy intensity substantially is equally problematical. For example, it has been estimated that to double the world's presently irrigated land would require an energy input equal to five percent of the world's petroleum *reserves* each year (Pimentel *et al.*, 1975).

Could industrialized agriculture, through food stockpiles, meet needs created by production shortfalls, while intermediate or small-scale energy subsidized technologies catch up with food requirements? Political and mathematical analyses of the necessary grain stocks have been undertaken (Eaton and Steele, 1976), but problems inherent in commodity food aid persist. There seems to be a strong view against regular food aid, which is said to allow recipient governments to focus on industrial development to the detriment of agriculture:

> To continue to allocate free or low cost food to governments that neglect their own rural areas . . . is counterproductive. It simply allows governments to put off the tedious and unglamorous task of helping their own people help themselves (Wortman, 1976:35).

Commodity aid may not solve food supply problems, since much food aid is sold by the recipient government and may not reach all needy persons. In addition, food aid may carry uneconomic "strings" such as the U.S. government's requirement that fifty percent of such shipments must be transported by American carriers, paid by recipient countries (Miyamoto, 1973). Program revenues loaned to private firms as part of policies to develop commercial opportunities may have negative repercussions. For example, American firms have used low interest government food aid funds to establish poultry industries in Colombia and South Korea, eventually diverting crop production from traditional foods, such as inexpensive dry beans as a protein source, to needs of the poultry industry (DeMarco and Sechler, 1975:46-50). Put simply, food aid has often been one dimension of continued economic imperialism by the industrialized nations.

Technology and technological research are indispensable for meeting the needs of the world food problem (Martin, 1975). Unfortunately, technology is so closely entwined in larger economic and political spheres that it is difficult to select technologies appropriate for local agricultural improvement which are free from the tentacles of the world energy economy and the world political system.

TABLE 28. TECHNOLOGICAL SOLUTIONS TO THE WORLD FOOD PROBLEM

Food Production:	Crop genetics
	Improved photosynthetic efficiency
	Disease resistance
	Fertilizer response
	Improved structure for light reception
	Improved nitrogen fixation
	New hybrid varieties
	New domestication
	Drought resistance
	Improved nutritional quality of crops
	Animal genetics
	Conversion efficiencies
	New domestication
	Farm operations
	Improved pest control
	New pesticides and agrichemicals
	Aerial application of pesticides
	Improved disease control
	Soil management improvements
	Mechanization
	Irrigation
	Factory farms
	Research
	Technology transfer
	Enhanced international research network
	Building agricultural research in developing areas
	Global monitoring systems
	Mathematical modeling
	Alternative food sources
	Protein synthesis
	Utilization of protein in oil crop residues
	Single cell protein
	Hydroponics
	Aquaculture
	Fishery management
	Fish meal
	Land resource development and preservation
	Arid areas
	Humid tropics
	Potential land in industrial areas
	Preservation of existing production areas
Food Provision:	Food preparation technology
	Nutrient fortification
	Food synthesis
	New processes to utilize existing crops
	Food stockpiles
	Emergency supplies
	Price stabilization
	Commodity food aid
	Food protection after harvest
	Agricultural specialization

Source: Compiled from references in the bibliography.

Population

The important role of population in the world food problem is illustrated by the agricultural portion of the World3 computer simulation model (Meadows *et al.*, 1974). World3 was created to explore the complex web of interrelationships among population, industry, agriculture, resources, and pollution. Each of the sections of the model is quasi-independent, with external connections to other sectors of the model. The agricultural sector (Figure 15) consists of several overlapping loops relating agricultural investments, resources, and food productivity (Randers and Zahn, 1974). Loop 1 is a negative feedback loop which adjusts food output from land in production to match food needs by the mechanism of investment in land development (Loop 1 = 1→2→3→4→5→1). Loop 2 (1→2→7→9→5→1) is similar in structure, adjusting food output by agricultural investment. However, such investment implies an additional loop (3 = 1→2→7→9→8→4→1), with positive feedback indicating that increased productivity will mean increased erosion, requiring further investment to increase productivity. Loop 4 is also a positive feedback loop (1→2→7→10→11→9→5→1) representing fertility degradation as a result of pollutants and other soil destroying processes. Fertility regeneration is possible in loop 5 (1→6→12→11→9→5→1) by allocation of land to restorative processes. Loop 6 (1→6→7→9→5→1), finally, suggests that food supplies can be augmented in the short run by bringing fallow land into production.

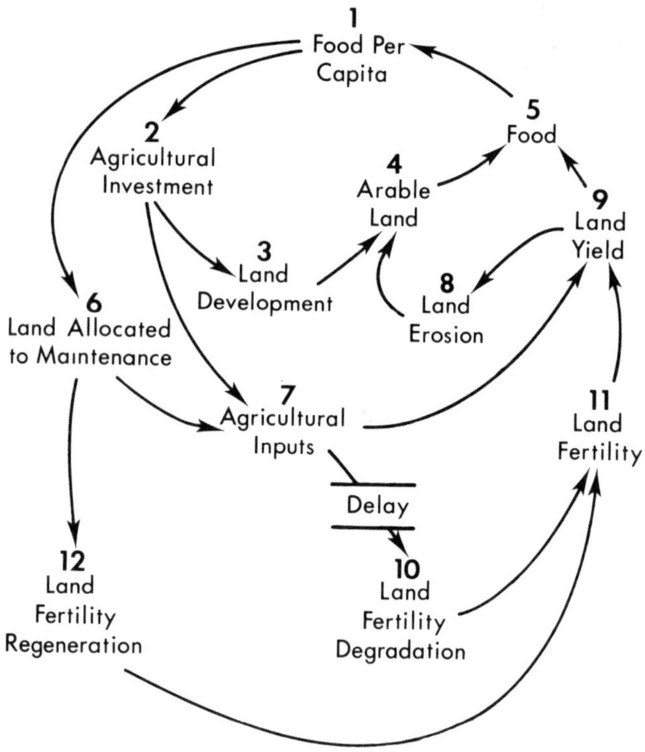

Figure 15. The Structure of the Agricultural Sector of the World3 Model. Adapted with permission from Meadows, Dennis L., William W. Behrens III, Donella H. Meadows, Roger F. Naill, Jørgen Randers, and Erich K. O. Zahn, *Dynamics of Growth in a Finite World*, copyright 1974 by Wright-Allen Press, Inc., Cambridge, Mass. 02142, p. 269.

The actual operation of the agricultural segment of World3 depends on a number of assumptions, expressed both verbally (Table 29) and mathematically. In addition, the mathematical equations describing the system must be calibrated for the "real world." The derived dynamics of the agricultural sector of World3 provide an interesting perspective on the role of population. Among a number of possible modes of population-resource interaction are those illustrated in Figure 16, and the question raised is which, if any, appears to operate in the agricultural sector. Tests of World3 included an historical calibration of the model, and a series of standard and optional runs to assess the model's sensitivity to estimation of parameters and technological innovations. Although there are many intriguing details at every step of the World3 model, we must limit mention here to two results of these tests. First, virtually any combination of calibrations and technological policy resulted in an overshoot and decline pattern of population and resources. The model suggests that the critical element that is altered by various parameter changes is the timing of the overshoot and decline, with each alternative simply accelerating or putting off the inevitable. Second, a series of equilibrium runs indicates that stability is possible, if the basic corrective measure—population stabilization—is taken sufficiently early. Figure 17 is the result of the run which assumes cessation of population growth in the year 2000.

One can debate whether 2000 is a firm date by which population growth must end. Indeed whether a leveling of population is more likely a cause or an effect of development is questionable (Frederiksen, 1969; Teitelbaum, 1975). Regardless, population numbers are a critical element in food system adequacy. Clearly population is crucial now given the unredressed imbalances in access to resources and differentials in food productivity that characterize the world food system. Even the most optimistic view of the success of various other "solutions" to the world food problem must be contingent on eventual balancing of population and resources. Whether this will be accomplished on a regional scale, with vastly different levels of material life, or on a world scale, with considerable equity in quality of life, is a choice that lies ahead.

Equity

The third solution cluster is one of equitable availability of resources to support food supply systems. Two elements dominate this cluster. The first is a decline in the energy intensiveness and resource demands of industrialized agriculture; the second, builds upon local initiative for agricultural development in nonindustrial areas. Realizing these goals may require more than humanitarian commitment: major segments of the world food system may have to be removed from the mechanisms of economic markets and from the domination of economics in the formulation of political policy. Related to this cluster are larger questions of world economic processes and access to resources and wealth; examination of food supply provides a brief glimpse of even broader questions of economic justice. As children, our mothers chided us for not eat-

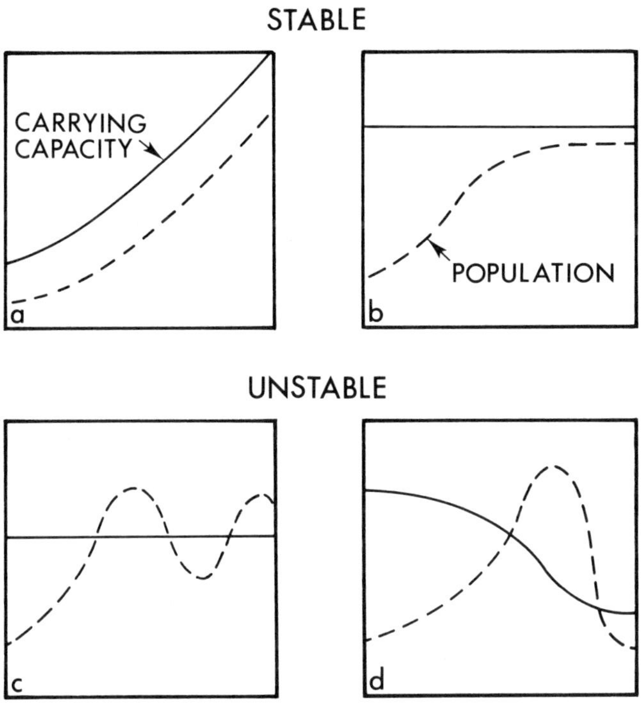

Figure 16. Population Growth and Resources. Four potential relationships between population and resources are illustrated: a) continuous growth; b) sigmoid or logistic approach to equilibrium; c) overshoot and oscillation; and d) overshoot and decline. Adapted with permission from Meadows, Dennis L., William W. Behrens III, Donella H. Meadows, Roger F. Naill, Jørgen Randers, and Erich K. O. Zahn, *Dyanmics of Growth in a Finite World*, copyright 1974 by Wright-Allen Press, Inc., Cambridge, Mass. 02142, p. 8.

ing while children elsewhere were starving. A mother's admonition rings frighteningly true.

Decreasing Resource Demands by Industrial Agriculture

The principal reason to alter the nature of industrialized agriculture is not, as skeptics have suggested, to redistribute food to hungry masses. Rather, the purpose is to make agricultural inputs accessible where their marginal returns are greater (and at lower prices); to improve the health of consumers in industrialized nations (by adjustment of diet quantity and quality to levels of necessity rather than gluttony); and to enhance the evolution of production technologies less vulnerable to the kinds of disruptions cited earlier in discussing the green revolution.

A variety of methods have been suggested for decreasing resource consumption in industrialized food supply systems (Brown, 1974:110-111; Wittwer, 1975:583). Energy inputs, for example, can be reduced by more efficient transportation systems, processing, and container technology. Irrigation efficiency can save water and legume rotations can save

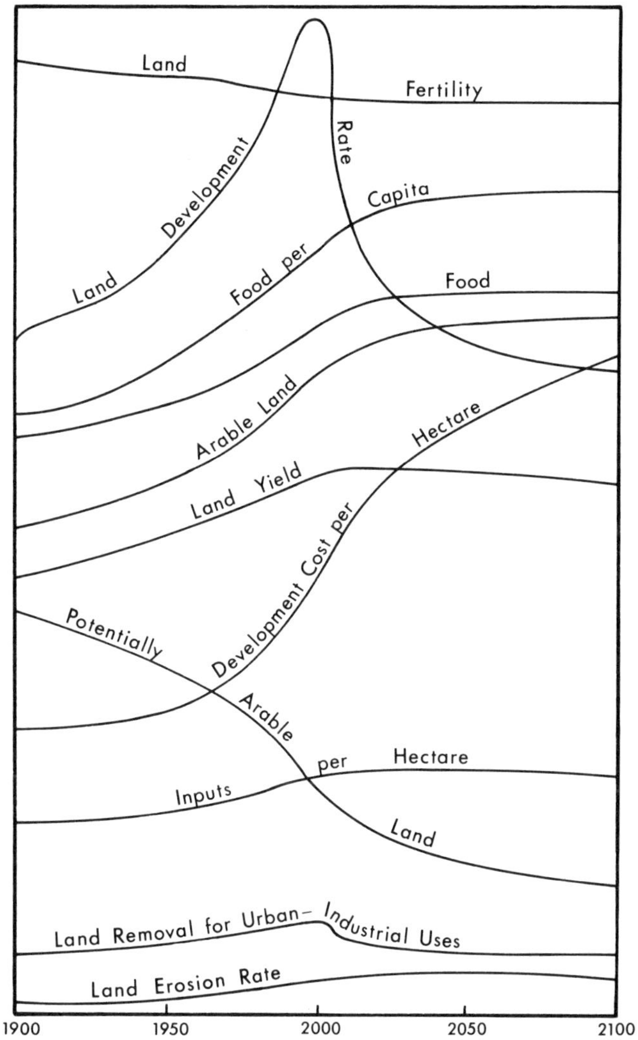

Figure 17. Stability in the Agricultural Sector of World3. This simulation assumes leveling of population, industrial output, and persistent pollution in the year 2000. Note the trajectory of each variable from 1900 to 2100. Adapted with permission from Meadows, Dennis L., William W. Behrens III, Donella H. Meadows, Roger F. Naill, Jørgen Randers, and Erich K. O. Zahn, *Dynamics of Growth in a Finite World*, copyright 1974 by Wright-Allen Press, Inc., Cambridge, Mass. 02142, p. 361.

fertilizer. Two specific examples will suggest that substantial levels of resource savings are possible with negligible threats to the provision of food supplies, and with both humanitarian and economic benefits. One example is the animal protein dependent diet; the other, use of agrichemicals.

Animal Versus Plant Protein

One of the major means for reducing the resource demands of industrialized agriculture is alteration of the animal based diet toward greater dependency on plant protein. This does not mean eliminating animal products from diets. As ruminants, cattle, sheep, and goats can use foods that cannot be digested by

TABLE 29. ASSUMPTIONS OF THE AGRICULTURAL SECTOR OF WORLD3

1. Food is produced from arable land and agricultural inputs (fertilizer, seed, pesticides).
2. Food output increases when the arable land area, the land fertility, or the amount of agricultural inputs are increased.
3. There are decreasing marginal returns to the use of agricultural inputs.
4. The amount of potentially arable land is finite, and development costs per hectare (for clearing, roads, irrigation dams) increase as the stock of potentially arable land decreases; in other words, the best and most accessible land is used first.
5. Newly developed land enters at the current *average* land fertility.
6. Arable land erodes irreversibly on a time scale of centuries when subject to intense cultivation, unless countermeasures are taken.
7. The stock of arable land is decreased by urban-industrial building activity, the rate of decrease depending on both population and industrial growth.
8. Total investment in agriculture increases in the long run with increasing industrial output per capita and in the short run when forced to do so by food shortages.
9. Agricultural investment can be used to develop new land or to increase the amount of agricultural inputs on present land. Investment is allocated on the basis of the relative marginal productivities of the options measured in vegetable-equivalent kilograms per dollar-year.
10. The capital intensive use of land can lead to persistent pollution of the land (high pesticide concentrations, salinity, heavy-metal poisoning).
11. Land fertility decreases on a time scale of decades when the level of persistent pollutants becomes high.
12. Land fertility regenerates itself over decades, and the process can be speeded up by proper land maintenance.
13. Farmers tend to maintain soil fertility by the proper use of capital except when pressured by extreme food shortages.
14. Land yield is reduced by air pollution.

Source: Assembled from Randers and Zahn (1974), in Meadows, *et al.*, *Dynamics of Growth in a Finite World.* Cambridge, Mass.: Wright-Allen Press, Inc., pp. 268-269. Copyright 1974 by Wright-Allen Press. Used by permission of D. L. Meadows and Wright-Allen Press.

man, including crop residues, forage crops produced on land not suited to other crops, and pasture that is not cultivable (Hodgson, 1976; Janick *et al.*, 1976). Forage crops as part of soil restoring rotations include nitrogen-fixing legumes. Thus, there are many cases where animals play an important role in food supply systems, where they are fully integrated into an ecologically rational rotation system, or where they indirectly harvest food resources that otherwise would be lost to man.

What can be reduced, however, is the allocation of feed grains to livestock, replacing that portion of animal protein intake with vegetable protein, or slackening protein intake to sufficient rather than superabundant levels. When fed with foods otherwise available to man, or food produced on land that could produce human food, great inefficiencies occur that can be attributed primarily to dietary predilections rather than to biological necessity. Various estimates have been made of the conversion efficiencies of livestock, as represented in the products consumed by humans (Table 30). The various efficiencies can be seen readily in market prices for food commodities, as well as in dietary advice for the poor—milk and eggs are much less expensive sources of animal protein than meats. Typically, meat prices range from broilers (least expensive), pork, and beef to lamb.

What would be gained by a decrease in the animal portion of diets? First, land for human food production would be increased greatly. Forage crop and feed grain land would be available for producing human food. Also, the amount of land needed to provide protein supplies would be vastly decreased (Table 31). Lockeretz (1975: 270) suggests that from three to six times as much land may be required for beef production compared to vegetable protein in

TABLE 30. ESTIMATED LIVESTOCK CONVERSION EFFICIENCIES

		Conversion Efficiency (%)[a]	
Animal	Product	Energy	Protein
Cow	Milk	17-19 (44)	25-31 (47)
	Beef	3-8	4-15
Pig	Pork	13-15	9-20
Sheep	Lamb	2-6	4-10
Chicken	Broilers	10-12	18-25
	Eggs	13-18 (20)	20-27 (36)
Goat	Milk	(25)	(44)

[a] Range of values includes typical values (Janick *et al.*, 1976; Van Vleck, 1975; Pimentel *et al.*; Heichel and Frink, 1975) and, in parentheses, potentially feasible values for milk and eggs (Byerly, 1967).

TABLE 31. COMPARATIVE RESOURCE USE FOR HUMAN PROTEIN SUPPLY ALTERNATIVES

Region and Production System[a]	Ratio of resource use for beef production to equivalent protein value in wheat/soybeans[b]			
	Cropland[c]	Energy[d]	Irrigation Water	Fertilizer
Texas High Plains				
Nonirrigated	3.0	10	—	8
Irrigated	2.1	5	4	19
Western Nebraska				
Nonirrigated	5.9	17	—	7
Irrigated crops	3.2	17	22	10
Irrigated crops and pasture	3.2	24	45	26
Northern Indiana	4.0	11	—	10
Georgia Coastal Plain				
Grain-fed	3.4	14	—	10
Pasture-fed	0.3	14	—	12

[a] Each beef production system includes pasture plus feed grain and protein supplement with feedlot finishing (except Georgia pasture-fed, which uses intensive pasture finishing).

[b] 1.0 pound lean beef versus 1.06 pounds whole wheat plus 0.23 pounds soybean. Vegetable protein sources irrigated where feed grains are irrigated.

[c] Not including pasture, some of which may be cultivable.

[d] Includes slaughter or flour processing.

Source: Lockeretz (1975:270). Used by permission of the *Journal of Soil and Water Conservation.*

most parts of the U.S. Second, water resource use would also be decreased, with irrigation requirements alone being four to 45 times greater for beef than for other field crops. Bradley (1962) suggests that the total water cost of beef is 25 times that of a vegetarian diet, considering water needs of plants supporting each system. Fertilizer requirements could be substantially decreased (one-seventh or less), as could be overall energy requirements (one-fifth or less). In summary, at 1970 U.S. crop yields, one hectare of cropped land could provide the annual protein needs of eight (wheat) to 16 (soybeans) adults, taking into account net utilizable protein (Pimentel et *al.*, 1975:756).

Several counterarguments are often raised concerning the utility of a potential decrease in animal consumption. First, it is pointed out that animals use some foods not available to man. However, an animal based diet in the industrialized world does represent a substantial allocation to livestock feed of both human foods and the resources to produce these foods. Second, it is argued that food resources made available by such a change could not efficiently or continuously be allocated to developing areas. This argument ignores, however, the decrease in resource demands such a change would bring—a decrease in demand for energy, fertilizers, and other agricultural inputs—thus providing less competition and perhaps lower prices for these inputs to the developing areas. Certainly only modest dietary change could reestablish the food stockpiles eroded during recent years. Furthermore, it is suggested that cultural, political, and economic obstacles make reducing meat output in America quite unlikely (Hopper, 1976:197). However, there are alternative processes to accomplish this end. One is a marked decrease in agrichemical use, driving meat product prices to commensurately higher levels, with increasing absolute differences in price between vegetable and animal sources of protein. Alternatively, the government could impose a ceiling on meat consumption, with coupon rationing and price controls imposed on animal products requiring resources that could produce nutritionally equivalent grains and legumes (Bäckstrand and Ingelstam, 1976). Finally, it is argued that scientific research is improving the efficiencies of feed conversion, and this, in conjunction with increased use of nonhuman food resources, will make animal derived foods available with more attractive costs and efficiencies than now exist. This argument, however, also suggests that the sooner potential human foods are denied for animal sustenance, the more rapidly these innovations will emerge, if for no other reason than price incentives to find innovative meat production technologies. We can conclude that what political decisions cannot now equitably and humanely determine about animal food consumption, the marketplace will eventually decide, too late perhaps to save squandered resources.

Agrichemicals

Pest eradication was a persistent theme accompanying use of chemical pesticides. In corn production, pesticide use increased thirty times in the last twenty years, but insect losses tripled. Pest resistance to chemicals, accompanied by high costs and environmental contamination, may change the pattern of pesticide use. Economic and environmental consequences of the eradication approach have lead to increasing emphasis on biological controls and pest management. Biological controls include those kinds of practices we mentioned in discussion of traditional agriculture—encouraging natural ecological balance by the nature of production—or use of natural parasites of pests. For example, California viticulturalists plant evergreen blackberry bushes near the vineyards as hosts for a wasp that preys on the grape leafhopper. Pest management includes the use of pesticides, but only on a "when needed" rather than a routine basis. Based on knowledge of development of the pest's life cycle, field inspections, and computer modeling, it is possible to predict when pest outbreaks are likely, and for farmers to receive advisory messages at that time. Potato producers in Pennsylvania, apple orchard owners in Michigan, and cotton producers in California can use integrated pest management systems. Biological controls and integrated pest control might break the vicious circle of increased pesticide use accompanied by rising crop losses as pests develop resistance (Brody, 1976).

It is conceivable that American farmers could eliminate the use of chemical fertilizers and pesticides. Chapman (1973), for example, suggested some consequences of regulation and prohibition of chemical use. Among these would be an increase in cultivated land with increased soil erosion as marginal land comes into production; higher food prices; smaller exportable surpluses; changed food appearance as a result of insect or bacterial spoilage; and potential decline in feedlots and other intensive production units dependent on chemicals. A return to crop rotation rather than monoculture would occur, as well as a decline in emigration from farming areas. Farm income could increase, a projection that is substantiated by studies of contemporary organic farms undertaken by the Center for the Biology of Natural Systems at Washington University (Lockeretz *et al.*, 1975). Sixteen organic farms in the Midwest were matched with control farms on the basis of size, soil type, location, and crop/livestock production system. Although both farm groups—organic and inorganic—were highly mechanized, the organic farms used only one-third the total energy input of the control farms per unit of production. The market value per acre was only slightly lower (eight percent) for the organic farms on an acreage basis, and the difference between crop value and operating costs was virtually identical for the two groups, because of the lower input costs for the organic farms. This conclusion suggests that if organic farming were widespread, and commodity prices increased accordingly, farm incomes could indeed increase.

It is difficult to measure the potential impact of pesticide prohibition. It is certainly erroneous to extrapolate from isolated organic farms amidst pesticide using neighbors. The neighbors provide an obvious spatial buffer decreasing the risk of pests, diseases, and weeds. Pimentel (1973) argued that prohibition of pesticide use in the U.S. would cause a seven percent increase in crop losses. This loss would be less than the annual ten percent production surplus. Pesticide prohibition would raise farm product value by one-quarter but retail food prices by only nine percent (Pimentel, 1973), since only one-third to two-fifths of food prices are determined by farm commodity prices (Chapman, 1973).

How could pesticide use be curtailed? One mechanism is price. At present, the benefit-cost ratio for pesticides is about three to one. This means that every dollar spent to purchase and apply pesticides yields three dollars increase in crop yield (Pimentel, 1973). If pesticide costs were increased, this ratio would be less favorable. Among mechanisms for raising these prices would be higher input costs for pesticide manufacturing (particularly petroleum prices); higher costs involved in proving safety of pesticides before use; and taxation. Increased prices might not be sufficient however, because decreased pesticide use accompanied by lower production would raise commodity prices, thus increasing the potential benefit-cost ratio. The alternative mechanisms are government control, such as exerted by the U.S. Environmental Protection Agency on environmental contamination and by the Food and Drug Administration on chemical residues on food, and consumer preference—a demand for organic foods. Outright legislated prohibition would face insurmountable political and industrial opposition.

A negative change in energy intensiveness in industrialized agriculture is a two-edged sword. On the positive side are these potential results:

- Impetus to maintain land in farms
- Increased income to farmers
- Greater farm employment
- Decreased ecological and health risk from chemical residues
- Use of natural pest controls
- Use of fertility enhancing crop rotations

On the other hand, these would also result:

- More land in cultivation
- Greater risk of erosion on marginal land
- Smaller crop production
- Higher food prices
- Less exportable surpluses

Alone, a simple decrease in energy use in industrialized agriculture—primarily through reduction in agrichemicals—would reduce significantly demand for fertilizers, pesticides, and herbicides, with the potential of making these commodities accessible at lower prices to developing areas where their use is appropriate. In other words, the means for production rather than the food itself would be reallocated.

The joint effects of dietary change and a decrease in energy intensity would be less ecologically disruptive, since the cropland saved from allocation to feedgrain production would offset decreased yields from agrichemical prohibition. Substantial economic dislocations would occur, since many American grain farmers also produce livestock products; agricultural input industries would be severely affected. Even if these two policies were adopted—a substantial decrease in animal-derived foods and a significant decline in agrichemical use—planning for cushioning the economic impacts is a major undertaking, in addition to the need for planning effective transfers to developing areas of mobilized resources:

> There are moral, ethical, environmental, and health arguments for the rich to alter their consumption to benefit the poor which do not need to be justified by the assumption that production possibilities are limited, or that food will be more costly and scarce . . . unless an effective institutional mechanism is established to transfer the sacrifices of the rich—in food, resources, or income—to the poor and malnourished, such sacrifices are not likely to be effective (USDA, 1974a:52).

Enhancing Local Initiative in Developing Areas

Increasingly it has been recognized that food supply systems of the developing areas are ultimately dependent on small-scale farmers and opportunities for increasing their productivity. Attention to the small farmer as opposed to large-scale schemes can be seen in statements of agricultural philosophy, policy, and practice. Ward (1974:25), for example, has suggested that:

> . . . it is upon (the) strategy of backing the small men—the half billion small farmers in the developing world—that the hopes of feeding most of mankind in the longer term depend.

Recent statements of intent by the World Bank (1976:15-17) parallel very closely new directions in USAID assistance, focusing on improving production in rural areas, with a strong component of equity or shared opportunity by significant numbers of rural poor:

> The bank believes that rural development, properly conceived and carried out, need not conflict with the objectives of higher food production. Indeed, studies indicate that small farmers are often more efficient in the use of farm resources than are large farmers. And though it may take longer to increase food output on small farms than on larger units—it is more difficult, for instance, to devise and implement development schemes involving large numbers of smallholders than those affecting only a few large-scale farmers—the Bank has concluded that, in the longer run, increases in food production of the magnitude required to satisfy worldwide demand can only be achieved by helping small farmers increase their productivity and output.

In practice, a number of studies have suggested the positive, focal role of small-scale farmers in development, such that enhancing their contributions to development planning is increasingly imperative (Knight, 1974; Morss *et al.*, 1975; Crosson, 1975). First, local farmers are often aware of problems and constraints on productivity. Second, ample evidence has accumulated to suggest that they respond positively to opportunities that are environmentally, economically, and socially sound, given appropriate material and institutional necessities. Third, much of the gains in productivity of cash and food crops in developing areas have been achieved by small-scale farmers. Finally, in traditional technologies, potential new ideas of wider applicability may be found.

One example of the potential use of traditional practices for agricultural development is the recent spread of water fern cultivation in rice paddies in Vietnam and southern China (Galston, 1975). In Thai Binh province of northern Vietnam, farmers have cultured the water fern, each year transplanting the fern to rice paddies. The fern vegetatively propagates, eventually covering the surface of the water. Rice yields are increased by fifty to 100 percent due to nitrogen fixation by a blue-green algae that lives in pockets on the water fern leaf. The symbiosis of the algae and fern is turned by these farmers into a symbiosis of rice and humans as well. Preserving the fern for the annual inoculation was a well-guarded, indeed valuable secret, kept even from local women who might marry into outside villages. The peasants were pursuaded to share this technology during the rice shortages of recent decades, and techniques of water fern culture are no longer secret. This traditional technology is proving of great value in the absence of fertilizers.

There is danger that many traditional technologies may be lost as "modern" education and agricultural practices supplant traditional knowledge and technology. If we view these traditional practices as a source of innovation, it is clear that they greatly enhance potentials for local agricultural improvement, as well as multiply the numbers of potential "solutions" to the world food problem. For example the Chinese government encourages traditional pest control practices along with modern procedures, with emphasis on prevention as well as on cure (Chiang, 1976).

Successful release of the latent potentials of farmers in the developing areas raises practical and ethical questions. Practically, science and technology must work cooperatively within local environments in seeking solutions. "Appropriate" or "alternative" technologies may make the technological progress of the industrialized world available at a scale relevant to the small holder in developing areas (Schumacher, 1973; Dickson, 1974; Wade, 1975b; Chidekel, 1975; Makhijani and Poole, 1975). However, is it ethically acceptable for donor nations to focus on development and diffusion of technologies that are now questioned in the industrialized areas? Can we justifiably encourage dependency on petroleum when we continue to squander this resource for our own food production and "necessities" of life? As Schumacher (1973:28) suggests:

> It is clear that the rich are in the process of stripping the world of its once-for-all endowment of relatively cheap and simple (to-use) fuels.

The vulnerability of green revolution technology

suggests the desirability of either fundamentally different approaches to agricultural development or a basic restructuring of the world economic system to assure careful husbanding of energy resources. Given the improbability of the latter, the former assumes crucial importance.

The Risk of Duality

The ultimate hope of technology is that somehow the gap between rich and poor will be closed for most of the world's people. A partial commitment to solutions that are locally adaptive and equitable—such as developing local initiative or assuring a modicum of access to resources—imposes a common risk with technological solutions that only partially succeed: increasing duality, and the potential of triage. Economic, social, and spatial dualities in developing areas are already familiar, as are similar distinctions between the industrialized and developing areas. From the acceptance of duality, triage is only a short step down a road toward the end of our morality. If we have neither the hope nor the moral commitment that all of humanity shall have equitable access to world resources and opportunity for freedom from hunger, we face an increasing probability of triage being seen as inevitable and therefore acceptable. For some, triage is a fashionable response to the threat imposed by equity and justice.

No short term palliative will redress present inequities in the world economic system. If prevention of hunger and famine are of highest priority in what must be a lengthy and difficult process, this at least can be accomplished now, with food production technologies presently available and resources courageously redirected. Berg (1975:35) has argued persuasively:

> After all the political and economic arguments, dealing with hunger and malnutrition is a moral issue—that demands a moral response. Why should it be so difficult to justify? We often hear that national policies should flow only from self-interest. We somehow have failed to recognize that doing good for the sake of doing good is self-interest. To most people, ethical concerns are of value. Compassion and human decency may not lend themselves neatly to cost-benefit analysis, but the desire for sound moral values, and the transfer of those values to future generations, is a legitimate rationale for Government action. Somehow, affluent societies must learn to accept this kind of self-interest as a basis for public policy.
>
> Given the facts, adherence to the lifeboat or triage theories is an intellectual and moral cop-out. To the extent that there is to be hunger and malnutrition, it will be a direct consequence of maldistribution of resources among and within nations. Enlightened policies and actions could prevent it, and we have no choice but to try. To do otherwise would reflect a fundamental and grievous change in the character of man. (Copyright © 1975 by the New York Times Company. Reprinted by permission.)

IV. PROSPECTS

. . . there is more involved in 'agricultural operations' than the production of incomes and the lowering of costs: what is involved is a society, the health, happiness, and harmony of man, as well as the beauty of his habitat.

E. F. Schumacher (1973:111-112)

In an article prepared prior to the U.N. World Food Conference in 1974, Barbara Ward characterized the world food problem as "the challenge of justice" (1974:25). Our analysis has led us to the same conclusion: the world food problem is our own creation, and solutions are available, if the industrialized nations are willing to take sufficient moral responsibility. Let us summarize the argument we have developed, share with you what would appear to us to be initial steps toward meeting immediate food needs, and finally suggest elements of a longer-term world food policy.

Although there is considerable uncertainty regarding the nature and extent of the world food problem, hunger is a contemporary reality, even within the United States. By the middle 1970's, much of the ability of the world food system to buffer fluctuations in food production had been eroded, with no immediate prospects for other than modest increases in productivity to match population growth, and substantial risk of famine in the absence of food stockpiles. North America assumed increasing importance as a food source for other areas, depending for production on technology with substantial economic, energetic, and resource subsidization.

Viewed from the perspective of the basic structural requisites of any food supply system, purported efficiencies of industrialized food production and provision are questionable, particularly as alternatives to traditional food supply systems in developing areas. Green revolution technologies assume the vulnerabilities of industrialized agriculture without its institutional features to protect against variability, having abandoned traditional means for coping with risk and uncertainty. The world food system is created by control over agricultural inputs and food commodities exerted by the industrialized nations. Commitment to elements of industrialized agriculture for food production links a developing nation to the world food system, and thus to the world food problem.

Although technological solutions are attractive, the world food problem is essentially moral or ethical rather than economic or technical in character. Food resources are inequitably distributed within and among nations. Changes in diet in the industrialized areas could immediately reestablish food stockpiles and in the longer view free resources for allocation to developing areas. Similarly, a decrease in energy intensity in industrialized agriculture could increase the availability of agricultural inputs for developing areas. Nevertheless, uncertainties of petroleum supply suggest caution in adoption of industrialized agriculture as a prototype for developing areas, even with "appropriate technologies." As Franke (1974:88) has observed in Java:

> The technology advocates, the rate-of-profit theorists, the military dictators, and the large landowners are attempting to produce enough food for the people of Java. They are failing. Their optimistic plans and programs have created only increased human suffering and promise more of the same. Perhaps solutions will come, not from the development experts, but from the small farmers and landless laborers of Java.

We find it particularly intriguing that one prescription for survival of the American industrialized agricultural system, based upon reduction of energy use, calls for smaller, less energy intensive farms; farm methods based on biological diversity; legumes to minimize fertilizer requirements; biological pest controls; solar crop drying; and windmills for irrigation energy (Clark, 1975). A potential convergence of industrialized agriculture and developing traditional cultivation systems is apparent.

Thus, solutions to the world food problem depend upon the willingness of the industrialized world to undergo modest "belt-tightening" in response to its own enlightened, moral self-interest. Given the need for population growth to stabilize with respect to resources and food production technologies, this belt-tightening may be insufficient to solve population problems, raising a larger "chicken and egg" proposition concerning the primacy of population growth or development, a topic not considered here.

Among potential policies to solve immediate hunger and famine problems are these:

(1) A voluntary decrease in calorie, protein, and particularly animal derived food consumption in the industrialized areas;

(2) Conversion of substantial feed grain cropland to human foods and subsequent establishment of food stockpiles and distribution

programs financed by industrialized economies; and

(3) The explicit assumption by the political leaders and citizens of the industrial nations of moral responsibility for meeting short-term world food needs.

In the long run, world food problems can only be addressed, in our view, as part of a fundamental reordering of world wealth. Among mechanisms of direct relevance to food supply are these:

(1) Food rationing in industrialized nations, with particular emphasis on limiting animal derived food intake dependent on resources capable of producing human food;
(2) Severe limitation on use of agricultural chemicals in industrial nations by rationing or taxation;
(3) Substantial energy taxation in the industrialized areas, with taxation inversely proportional to reserves of each fuel source and with small tax rebates to cover true energy consumption necessities by the poor; and
(4) A formal commitment to development financing by the industrialized nations, each of which is assessed proportionally to gross national product and to indices of consumption of world resources.

Perhaps most important is a policy that cannot be legislated, a policy of moral commitment to equitable, humane solutions to the world food problem. We must recognize that these solutions will have a profound impact on ourselves as well as on the world's poor. Adjusting to these impacts may be one of the greatest challenges those accustomed to the material consumption standards of the industrialized world will face. Nevertheless, a commitment must be made. As persuasive and as seemingly dispassionate as scientific analysis of the situation might appear, arguments about solutions to the world food problem will remain inherently philosophical, moral, and political. To solve the world food problem, we will pay the necessary price, if only because triage is unconscionable. If economics should have any bearing upon our decision, perhaps we could take comfort that moral action will be "cheaper" while energy supplies are still relatively plentiful.

> I sit on a man's back, choking him, and making him carry me, and yet assure myself and others that I am very sorry for him and wish to ease his lot by any means possible, except getting off his back (Leo Tolstoy, quoted by Clinton, 1976).

BIBLIOGRAPHY

Allan, W., 1965, *The African Husbandman*. Edinburgh: Oliver and Boyd.

Andrilenas, P. A., 1975, *Farmers' Use of Pesticides in 1971 . . . Extent of Crop Use*. Washington: USDA Economic Research Service, Agricultural Economic Report 268.

Atkinson, L. J. and D. E. Kunkel, 1976, *High Yielding Varieties of Rice in the Philippines*. Washington: USDA Economic Research Service, Foreign Agricultural Economic Report 113.

Bäckstrand, G. and L. Ingelstam, 1976, "We Could Remodel Sweden," *Current Sweden*, Vol. 76/1.

Bajema, C. J., 1972, "Transmission of Information About the Environment in the Human Species: A Cybernetic View of Genetic and Cultural Evolution," *Social Biology*, Vol. 19, No. 3, pp. 224-226.

Barnet, R. J., 1976. "Multinationals: A Dissenting View," *Saturday Review*, Vol. 3, No. 9, pp. 11, 58.

Barnet, R. J. and R. E. Muller, 1975, *Global Reach: The Power of Multinational Corporations*. New York: Simon and Schuster.

Bennett, M. K., 1954, *The World's Food*. New York: Harper.

Berg, A., 1975, "The Trouble with Triage," *New York Times Magazine*, June 15, 1975, pp. 26-35.

Berlin, B. *et al.*, 1974, *Principles of Tzeltal Plant Classification: An Introduction to the Botanical Ethnography of a Mayan-Speaking People of Highland Chiapas*. New York: Academic Press.

Boerma, A. H., 1970, "A World Agricultural Plan," *Scientific American*, Vol. 223, No. 2, pp. 54-69.

Borchert, J. R., 1971, "The Dust Bowl in the 1970's," *Annals of the Association of American Geographers*, Vol. 61, pp. 1-22.

Boserup, E., 1965, *The Conditions of Agricultural Growth*. Chicago: Aldine.

Bradley, C. C., 1962, "Human Water Needs and Water Use in America," *Science*, Vol. 138, pp. 489-491.

Brody, J. E., 1976, "Farmers Turn to Pest Control in Place of Eradication," *New York Times*, August 1, 1976.

Brooke, C., 1967, "Types of Food Shortages in Tanzania," *Geographical Review*, Vol. 57, pp. 333-357.

Brown, L. R. with E. P. Eckholm, 1974, *By Bread Alone*. New York: Praeger.

Brown, L. R., 1975, "The World Food Prospect," *Science*, Vol. 190, pp. 1053-1059.

Brown, N. L. and E. R. Pariser, 1975, "Food Science in Developing Countries," *Science*, Vol. 188, pp. 589-593.

Bryson, R. A., 1976, "The Lessons of Climatic History," *Ecologist*, Vol. 6, pp. 205-211.

Buckley, W., 1968, "Society as a Complex Adaptive System," in W. Buckley (ed.), *Modern Systems Research for the Behavioral Scientist*. Chicago: Aldine.

Byerly, T. C., 1967, "The Efficiency of Feed Conversion," *Science*, Vol. 157, pp. 890-895.

Cervinka, V. *et al.*, 1974, *Energy Requirements for Agriculture in California*. Sacramento: California Department of Food and Agriculture, in cooperation with University of California, Davis.

Chancellor, W. J. and J. R. Goss, 1976, "Balancing Energy and Food Production, 1975-2000," *Science*, Vol. 192, pp. 213-218.

Chang, J.-H., 1968, "The Agricultural Potential of the Humid Tropics," *Geographical Review*, Vol. 58, pp. 333-361.

________________, 1970, "Potential Photosynthesis and Crop Productivity," *Annals of the Association of American Geographers*, Vol. 60, pp. 92-101.

Chapman, D., 1973, "An End to Chemical Farming?" *Environment*, Vol. 15, No. 2, pp. 12-17.

Chiang, H. C., 1976, "Pest Control in the People's Republic of China," *Science*, Vol. 192, pp. 675-676.

Chidekel, D., 1975, " 'Small is Beautiful' as a Book and a Bum Steer," *Science for the People*, Vol. 7, pp. 17-19.

Citizens Board of Inquiry into Hunger and Malnutrition in the U.S., 1968, *Hunger USA: A Report*. Boston: Beacon Press.

Clark, C., 1967, *Population Growth and Land Use*. London: Macmillan.

Clark, W., 1975, "U.S. Agriculture is Growing Trouble as Well as Crops," *Smithsonian*, Vol. 5, No. 10, pp. 59-65.

Cleaver, H. M., Jr., 1972, "The Contradiction of the Green Revolution," *Monthly Review*, Vol. 24, pp. 80-111.

Clinton, R. L., 1976, "The Specter of Starvation," *War on Hunger*, Vol. 10, No. 10, pp. 10-14.

Cochrane, W. W., 1969, *The World Food Problem*. New York: Crowell.

Commoner, B. *et al.*, 1974, *The Effect of Recent Energy Price Increases on Field Crop Production Costs*. St. Louis: Washington University, Center for the Biology of Natural Systems.

Conklin, H. C., 1967, "Some Aspects of Ethnographic Research in Ifugao," *Transactions of the New York Academy of Sciences*, Series II, Vol. 30, pp. 99-121.

Connell, K. H., 1950, *The Population of Ireland*. Oxford: Clarendon.

Cook, E., 1976, *Man, Energy and Society*. San Francisco: W. H. Freeman.

Crosson, P. R., 1975, "Institutional Obstacles to Expansion of World Food Production," *Science*, Vol. 188, pp. 519-523.

Dalrymple, D. G., 1974, *Development and Spread of High Yielding Varieties of Wheat and Rice in the Less Developed Nations*. Washington: USDA Economic Research Service, Foreign Agricultural Economic Report 95.

Dalrymple, D. G., 1975, *Measuring the Green Revolution: The Impact of Research on Wheat and Rice Production*. Washington: USDA Economic Research Service, Foreign Agricultural Economic Report 106.

Damon, P. E. and S. M. Kunen, 1976, "Global Cooling?" *Science*, Vol. 193, pp. 447-453.

Dasmann, R. *et al.*, 1973, *Ecological Principles for*

Economic Development. London: Wiley.
de Alcantara, C. H., 1973-74, "The Green Revolution as History: The Mexican Experience," *Development and Change,* Vol. 5, pp. 25-44
de Castro, J., 1952, *The Geography of Hunger.* Boston: Little, Brown.
DeMarco, S. and S. Sechler, 1975, *The Fields Have Turned Brown.* Washington: Agribusiness Accountability Project.
deWit, C. T., 1967, "Photosynthesis: Its Relationship to Overpopulation," in San Pietro *et al., Harvesting the Sun.* New York: Academic Press.
Dickson, D., 1974, *Alternative Technology and the Politics of Technical Change.* Glasgow: Fontana.
Duckham, A. N., and G. B. Masefield, 1970, *Farming Systems of the World.* New York: Praeger.
Dumont, R. and B. Rosier, 1969, *The Hungry Future.* New York: Praeger.
Dwyer, J. T. and J. Mayer, 1975, "Beyond Economics and Nutrition: The Complex Basis of Food Policy," *Science,* Vol. 188, pp. 566-570.
Eaton, D. J. and W. S. Steele, 1976, *Analysis of Grain Reserves, A Proceedings.* Washington: USDA Economic Research Service, Report 634.
Eckholm, E. P., 1976, *Losing Ground: Environmental Stress and World Food Prospects.* New York: W. W. Norton.
Ehrlich, P. R., 1968, *The Population Bomb.* New York: Ballantine Brooks.
Evans, C. D. and R. W. Simunek, 1976, *Balance Sheet of the Farming Sector, 1975.* Washington: USDA Economic Research Service, Agriculture Information Bulletin 389.
FAO, 1946, *World Food Survey.* Washington: FAO.
——————, 1952, *Second World Food Survey.* Rome: FAO.
——————, 1961, *Development Through Food, A Strategy for Surplus Utilization.* Rome: FAO.
——————, 1963, *Third World Food Survey.* Rome: FAO
——————, 1969, *State of Food and Agriculture.* Rome: FAO.
——————, 1971a, *Food Balance Sheets, 1964-1966 Average.* Rome: FAO.
——————,,1971b, *Agricultural Commodity Projections, 1970-1980.* Rome: FAO.
——————, 1974, "Population, Food Supply and Agricultural Development," *Monthly Bulletin of Agricultural Economics and Statistics,* Vol. 23, No. 9, pp. 1-13.
——————, 1975a, *Production Yearbook,* Vol. 28, 1974. Rome: FAO.
——————, 1975b, "World Food and Agricultural Situation—February 1975," *Monthly Bulletin of Agricultural and Economic Statistics,* Vol. 24, No. 4, pp. 1-19.
Farvar, M. T. and J. P. Milton, 1972, *The Careless Technology: Ecology and International Development.* Garden City, N.Y.: Natural History Press.
Feder, E., 1973-74, "Six Plausible Theses About the Peasant's Perspectives in the Developing World," *Development and Change,* Vol. 5, pp. 1-24.
Firey, W., 1960, *Man, Mind and Land.* Glencoe, Ill.: The Free Press.
Franke, R. W., 1974, "Miracle Seeds and Shattered Dreams in Java," *Natural History,* Vol. 83, No. 1, pp. 10-18, 84-88.
Frankel, O. H., 1973, *Survey of Genetic Resources in Their Centers of Diversity.* Rome: FAO and International Biological Programme.
Frederiksen, H., 1969, "Feedbacks in Economic and Demographic Transition," *Science,* Vol. 166, pp. 837-847.
Galston, A. W., 1975, "The Water Fern-Rice Connection," *Natural History,* Vol. 84, No. 10, pp. 10-11.
Geertz, C., 1963, *Agricultural Involution: The Process of Ecological Change in Indonesia.* Berkeley: University of California Press.
Green, M. J. *et al.,* 1972, *The Closed Enterprise System.* New York: Bantam.
Griffin, K., 1972, *The Green Revolution: An Economic Analysis.* Geneva: United Nations Research Institute for Social Development, Report 72.6.
Grigg, D. B., 1974, *The Agricultural Systems of the World.* New York: Cambridge University Press.
Harlan, J. R., 1975, "Our Vanishing Genetic Resources," *Science,* Vol. 188, pp. 618-621.
Harris, M., 1966, "The Cultural Ecology of India's Sacred Cattle," *Current Anthropology,* Vol. 7, pp. 51-66.
——————, 1972, "One Man's Food is Another Man's Whitewash," *Natural History,* Vol. 81, No. 9, pp. 12, 14.
Heady, E. O., 1976, "The Agriculture of the U.S.," *Scientific American,* Vol. 235, No. 3, pp. 106-127.
Heichel, G. H., 1973, *Comparative Efficiency of Energy Use in Crop Production.* New Haven: Connecticut Agricultural Experiment Station, Bulletin 739.
——————, 1974a, "Energy Needs and Food Yields," *Technology Review,* Vol. 76, pp. 18-25.
——————, 1974b, *Auxiliary Energy Requirements and Food Energy Yields of Selected Food Crops.* New Haven: Connecticut Agricultural Experiment Station, Special Soils Bulletin 36.
Heichel, G. H. and C. R. Frink, 1975, "Anticipating the Energy Needs of American Agriculture," *Journal of Soil and Water Conservation,* Vol. 30, pp. 48-53.
Heilbroner, R. L., 1963, *The Great Ascent.* New York: Harper and Row.
Higbee, E., 1963, *Farms and Farmers in an Urban Age.* New York: Twentieth Century Fund.
Hightower, J., 1972, *Hard Tomatoes, Hard Times.* Washington: Agribusiness Accountability Project.
——————, 1973, "Agribusiness and Agrigovernment: Power, Profits, and Poverty," in G. L. Mason and F. Velter (eds.), *The Politics of Exploitation.* New York: Random House.
——————, 1975, *Eat Your Heart Out, Food Profiteering in America.* New York: Crown Publishers.
Hirst, E., 1974, "Food-Related Energy Requirements," *Science,* Vol. 184, pp. 134-138.
Hodgson, H. J., 1976, "Forage Crops," *Scientific American,* Vol. 234, No. 2, pp. 61-75.
Hopper, W. D., 1976, "The Development of Agriculture in Developing Countries," *Scientific American,* Vol. 235, No. 3, pp. 196-205.
International Rice Research Institute, 1975, *Changes in Rice Farming in Selected Areas of Asia.* Los Banos, Philippines: IRRI.
Janick, J., C. H. Noller, and C. L. Rhykerd, 1976, "The Cycles of Plant and Animal Nutrition," *Scientific American,* Vol. 235, No. 3, pp. 74-86.
Janzen, D. H., 1973, "Tropical Agroecosystems," *Science,* Vol. 182, pp. 1212-1219.
Jenks, G. F. and W. M. Kollmorgen, 1958, "Suitcase Farm-

ing in Sully County, South Dakota," *Annals of the Association of American Geographers*, Vol. 48, pp. 27-40.

Jennings, P. R., 1976, "The Amplification of Agricultural Production," *Scientific American*, Vol. 235, No. 3, pp. 180-194.

Knight, C. G., 1974, *Ecology and Change: Rural Modernization in an African Community*. New York: Academic Press.

Krenz, R. *et al.*, 1976, *Costs of Producing Food Grains, Feed Grains, Oilseeds, and Cotton, 1974-76*. Washington: USDA Economic Research Service, Agricultural Economic Report 338.

Lakshminarayana, V. and P. K. Menon, 1972, "Exposure of the General Population in Hyderabad to DDT," *Journal of Food Science and Technology*, Vol. 9, p. 82.

Lappé, F. M., 1971, *Diet for a Small Planet*. New York: Friends of the Earth/Ballantine.

Leach, G., 1975, *Energy and Food Production*. London: International Institute for Environment and Development.

Leon, B., 1975, "Agriculture: A Sacred Cow," *Environment*, Vol. 17, No. 9, pp. 38-40.

Lerzu, C., 1974, "The New Food Chain," *Environmental Action*, Vol. 5, No. 20, pp. 7-10.

Lockeretz, W., 1975, "Consumption of Agricultural Resources in the Production of Fat Cattle and Food Grains," *Journal of Soil and Water Conservation*, Vol. 30, pp. 268-271.

Lockeretz, W. *et al.*, 1975, *A Comparison of the Production, Economic Returns, and Energy Intensiveness of Corn Belt Farms That Do and Do Not Use Inorganic Fertilizers and Pesticides*. St. Louis: Washington University, Center for the Biology of Natural Systems.

Loomis, R. S., 1976, "Agricultural Systems," *Scientific American*, Vol. 235, pp. 98-105.

Madden, J. P., 1967, *Economies of Size in Farming*. Washington: USDA Economic Research Service, Agricultural Economic Report 107.

Makhijani, A. and A. Poole, 1975, *Energy and Agriculture in the Third World*. Boston: Ballinger.

Malassis, L., 1976, *The Rural World: Education and Development*. Paris: UNESCO.

Martin, W. P., 1975, *All-Out Food Production: Strategy and Resource Implications*. Madison, Wisc.: American Society of Agronomy, Special Publication 23.

Mascarenhas, A. C. *et al.*, 1973, "Studies in Famines and Food Shortages," Special Number, *Journal of the Geographical Association of Tanzania*, Vol. 8.

May, J. M., 1974, "The Geography of Nutrition," in J. M. Hunter (ed.), *The Geography of Health and Disease*. Chapel Hill: University of North Carolina, Department of Geography, Studies in Geography No. 6.

Mayer, J., 1975, "Management of Famine Relief," *Science*, Vol. 188, pp. 571-577.

________________, 1976, "The Dimensions of Human Hunger," *Scientific American*, Vol. 235, No. 3, pp. 40-49.

Meadows, D. H. *et al.*, 1972, *The Limits to Growth*. New York: Universe Books.

Meadows, D. L. *et al.*, 1974, *Dynamics of Growth in a Finite World*. Cambridge, Mass.: Wright-Allen Press, Inc.

Miller, J., 1973, "Genetic Erosion: Crop Plants Threatened by Government Neglect," *Science*, Vol. 182, p. 1231.

Miyamoto, I., 1973, "The Real Value of Tied Aid: The Case of Indonesia 1967-69," *Economic Development and Cultural Change*, Vol. 22, pp. 436-452.

Morgan, W. B. and R. P. Moss, 1970, "Soils, Plants, and Farmers in West Africa," in J. P. Garlick and R. W. J. Keay (eds.), *Human Ecology in the Tropics*. Oxford: Pergamon.

Morss, E. R. *et al.*, 1975, *Strategies for Small Farmer Development: An Empirical Study of Rural Development Projects*. Washington: Development Alternatives, Inc., for U.S. Agency for International Development.

OECD, 1973, *Food Consumption Statistics 1955-1971*. Paris: Organization for Economic Cooperation and Development.

Oldfield, M., 1976, *The Utilization and Conservation of Genetic Resources*, unpublished M.A. thesis, The Pennsylvania State University.

Ormrod, R. K., 1974, *Adaptation in Cultural Ecosystems*, unpublished Ph.D. dissertation, The Pennsylvania State University.

Paddock, W. and P. Paddock, 1964, *Hungry Nations*. Boston: Little and Brown.

________________, 1967, *Famine–1975! America's Decision: Who Will Survive?* Boston: Little and Brown.

Palmer, I., 1972, *Science and Agricultural Production*. Geneva: United Nations Research Institute for Social Development, Report 72.8.

Pennsylvania Farm Management and Business Analysis Service, 1976, *1975 FMBA Service Farm Business Summary*. Camp Hill, Pa.: Pennsylvania Farmer's Association.

Perelman, M. and K. P. Shea, 1972, "The Big Farm," *Environment*, Vol. 14, No. 10, pp. 10-15.

Périssé, J., F. Sizaret, and P. Francois, 1969, cited in WHO, 1973, *Energy and Protein Requirements*. Geneva: WHO Technical Report Series 522, p. 20.

Pimentel, D., 1973, "Realities of a Pesticide Ban," *Environment*, Vol. 15, No. 2, pp. 18-31.

Pimentel, D. *et al.*, 1973, "Food Production and the Energy Crisis," *Science*, Vol. 182, pp. 443-449.

________________, 1974, *Workshop on Research Methodologies for Studies of Energy, Food, Man and Environment, Phase 1*. Ithaca: Cornell University, Center for Environmental Quality Management.

________________, 1975, "Energy and Land Constraints in Food Protein Production," *Science*, Vol. 190, pp. 754-761.

Poleman, T. T., 1975, "World Food: A Perspective," *Science*, Vol. 188, pp. 510-518.

Polanyi, K. *et al.*, 1957, *Trade and Market in the Early Empires*. New York: Free Press.

Porter, P. W., 1965, "Environmental Potentials and Economic Opportunities: A Background for Cultural Adaptation," *American Anthropologist*, Vol. 67, pp. 409-20.

Porter, P. W. and A. de Souza, 1974, *The Underdevelopment and Modernization of the Third World*. Washington: Association of American Geographers, Commission on College Geography, Resource Paper 28.

Prabowo, D. and Sajogyo, 1975, "Sidoarjo, East Java and Subang, West Java," in IRRI, *Changes in Rice Farming in Selected Areas of Asia*. Los Banos, Philippines: IRRI.

President's Science Advisory Committee, 1967, *The World Food Problem*. Washington: U.S. Government Printing Office.

Ramparts, 1971, "Hard Times: The Agri-Welfare Roll," Vol. 10, No. 4, pp. 10-11.

Randers, J. and E. K. O. Zahn, 1974, "Agriculture Sector," in Meadows *et al.*, *Dynamics of Growth in a Finite World*. Cambridge, Mass.: Wright-Allen Press, Inc., pp. 257-368.
Rappaport, R. A., 1971, "The Flow of Energy in an Agricultural Society," *Scientific American*, Vol. 225, No. 3, pp. 116-132.
Rao, V. K. R. V., 1974, *Growth With Justice in Asian Agriculture*. Geneva: United Nations Research Institute of Social Development, Report 74.2.
Rasmussen, W. D., G. L. Baker, and J. S. Ward, 1976, *A Short History of Agricultural Adjustment*. Washington: USDA Economic Research Service, Agriculture Information Bulletin 391.
Russell, J., 1954, *World Population and World Food Supplies*. London: Allen and Unwin.
Ruthenberg, H., 1971, *Farming Systems in the Tropics*. Oxford: Clarendon.
Sanderson, F. H., 1975, "The Great Food Fumble," *Science*, Vol. 188, pp. 503-509.
San Pietro, A., F. A. Greer, and T. J. Army, 1967, *Harvesting the Sun*. New York: Academic Press.
Saxe, J. G., 1882, *The Poetical Works of John Godfrey Saxe*. Boston: Houghton Mifflin.
Schultze, C. L., 1971, *The Distribution of Farm Subsidies: Who Gets the Benefit?* Washington: Brookings Institute.
Schumacher, E. F., 1973, *Small is Beautiful: Economics as if People Mattered*. New York: Harper & Row.
Science, 1975, Special Issue, "Food," Vol. 188, pp. 501-653.
Scientific American, 1976, Special Issue, "Food and Agriculture," Vol. 235, No. 3.
Scrimshaw, N. S. and V. R. Young, 1976, "The Requirements of Human Nutrition," *Scientific American*, Vol. 235, No. 3, pp. 50-64.
Simoons, F. J., 1967, *Eat Not This Flesh*. Madison: University of Wisconsin Press.
Sommers, C. E., 1976, "Crop Management," *Successful Farming*, Vol. 74, No. 2, p. 18.
Spencer, J. E. and R. J. Horvath, 1963, "How Does An Agricultural Region Originate?" *Annals of the Association of American Geographers*, Vol. 53, pp. 74-92.
Sprout, H. and M. Sprout, 1974, *Multiple Vulnerabilities: The Context of Environmental Repair*. Princeton, N.J.: Princeton University, Center of International Studies, Research Monograph 40.
Steinhart, J. S. and C. E. Steinhart, 1974, "Energy Use in the U.S. Food System," *Science*, Vol. 184, pp. 307-316. This article is modified from C. E. Steinhart and J. S. Steinhart, 1974, *Energy: Sources, Uses, and Role in Human Affairs*. North Scituate, Mass.: Duxbury Press.
Successful Farming, 1976, "Assessing Farmland: High Priced Land Gets a Break," Vol. 74, No. 9, p. 7.
Szentes, T., 1971, *The Political Economy of Underdevelopment*. Budapest: Akademiai Kiado.
Teitelbaum, M. S., 1975, "Relevance of Demographic Transition Theory for Developing Countries," *Science*, Vol. 188, pp. 420-425.
Thomas, I. D., 1972, "Infant Mortality in Tanzania," *East African Geographical Review*, Vol. 10, pp. 5-26.
Thornthwaite, C. W., 1941, *Atlas of Climatic Types in the United States*. Washington: U.S. Government Printing Office.
Union for Radical Political Economics, 1973, *Fact Sheets*. New York: URPE.
USAID, 1974, *Voluntary Foreign Aid Programs*. Washington: AID Bureau for Population and Humanitarian Assistance, Office of Private and Voluntary Cooperation.
__________, 1975a, *AID in An Interdependent World, A Summary of the Presentation to Congress, Foreign Assistance Programs for Fiscal 1976*. Washington: Agency for International Development.
__________, 1975b, *U.S. Overseas Loans and Grants and Assistance from International Organizations*. Washington: Agency for International Development, Office of Finanical Management.
U.S. Bureau of the Census, 1967, *1967 Census of Manufactures, Concentration Ratios in Manufacturing*. Washington: Department of Commerce, Bureau of the Census.
USDA, 1956, *Household Food Consumption Survey, 1955*. Washington: USDA Agricultural Research Service.
__________, 1958. *A Chronology of American Agriculture, 1790-1956*. Washington: USDA Agricultural Marketing Service.
__________, 1961, *The World Food Budget 1962 and 1966*. Washington: USDA Economic Research Service, Foreign Agricultural Economic Report 4.
__________, 1964, *The World Food Budget 1970*. Washington: USDA Economic Research Service, Foreign Agricultural Economic Report 19.
__________, 1969, *Household Food Consumption Survey 1965-66*. Washington: USDA Agricultural Research Service.
__________, 1970, *World Demand Prospects for Agricultural Exports of Less Developed Countries in 1980*. Washington: USDA Economic Research Service, Foreign Agricultural Economic Report 60.
__________, 1971, *Selected U.S. Crop Budgets*. Washington: USDA Economic Research Service.
__________, 1972a, *Usual Planting and Harvesting Dates*. Washington: USDA Statistical Reporting Service, Agriculture Handbook 283.
__________, 1972b, *The Market Structure of the Food Industry*. Washington: USDA Market Economics Division, Market Research Report 971.
__________, 1974a, *The World Food Situation and Prospects to 1985*. Washington: USDA Economic Research Service, Foreign Agricultural Economic Report 98.
__________, 1974b, *Our Land and Water Resources*. Washington: USDA Economic Research Service, Miscellaneous Publication 1290.
__________, 1975a, *That We May Eat, The Yearbook of Agriculture 1975*. Washington: U.S. Government Printing Office.
__________, 1975b, *1975 Changes in Farm Production and Efficiency, A Summary Report*. Washington: USDA Economic Research Service, Statistical Bulletin 548.
__________, 1976, *Firm Enterprise Data System*. Stillwater, Okla. USDA Economic Research Service, Commodity Economics Division and Oklahoma State University.
USFEA, 1976, *Energy Use in the Food System*. Washington: U.S. Federal Energy Administration.
U.S. Senate Select Committee on Nutrition and Human Needs, 1973, *Hunger 1973*. Washington: U.S. Government Printing Office.
__________, 1974, *Nutrition and Food Availa-*

bility. Washington: U.S. Government Printing Office.
U.S. Senate Subcommittee on Monopoly, 1968, *Corporation Farming: Great Plains and Upper Midwest Region.* Washington: Government Printing Office.
________________, 1972, *Role of Giant Corporations: Corporate Secrecy: Agribusiness:* Washington: U.S. Government Printing Office.
Van Vleck, G., 1975, "For Human Nutrition: Grain vs. Meat," *The Professional Nutritionist,* Vol. 7, No. 2, pp. 1, 6-10.
Vogt, W., 1948, *Road to Survival.* New York: Sloane.
Wade, N., 1975a, "International Agricultural Research," *Science,* Vol. 188, pp. 585-589.
________________, 1975b, "New Alchemy Institute: Search for an Alternative Agriculture," *Science,* Vol. 187, pp. 727-729.
Walters, H., 1975, "Difficult Issues Underlying Food Problems," *Science,* Vol. 188, pp. 524-530.
War on Hunger, 1976, Special issue on private and voluntary agencies, Vol. 10, No. 7.
Ward, B., 1974, "The Fat Years and the Lean," *The Economist,* Vol. 253, No. 6845, pp. 19-25.
Webb, W. P. 1931, *The Great Plains.* New York: Ginn and Company.
Whittaker, R. H. and G. Likens, 1975. "The Biosphere and Man," in H. Lieth and R. H. Whittaker (eds.), *Primary Productivity of the Biosphere.* New York: Springer-Verlag.
Wittwer, S. H., 1975, "Food Production: Technology and the Resource Base," *Science,* Vol. 188, pp. 597-584.
World Bank, 1976, *Annual Report 1976.* Washington: World Bank Group.
World Health Organization, 1973, *Energy and Protein Requirements.* Geneva: WHO Technical Report Series 522.
Wortman, S., 1976, "Food and Agriculture," *Scientific American,* Vol. 235, No. 3, pp. 30-39.
Zeimetz, K. A. *et al.,* 1976, *Dynamics of Land Use in Fast Growth Areas.* Washington: USDA Economic Research Service, Agricultural Economic Report 325.